BIOLOGY REVISIONED

Willis W. Harman, Ph.D.
& Elisabet Sahtouris, Ph.D.

North Atlantic Books
Berkeley, California

Biology Revisioned

Published by
North Atlantic Books
P.O. Box 12327, Berkeley, California 94712

Cover art by Ruth Terrill
Cover and book design by Nancy Koerner
Illustrations on pages 4, 5, 19, 32, 49, 70, 71, 93, 105 by Adrienne Smucker
Cartoon on page 7 by Carol Guion
Hubble Telescope photo on page 120 courtesy Jeff Hester & Paul Scowen
(Arizona State University), and NASA

Printed in the United States of America

Biology Revisioned is co-sponsored by the Society for the Study of Native Arts and Sciences and the Institute of Noetic Sciences.

The Society for the Study of Native Arts and Sciences is a nonprofit educational corporation whose goals are to develop an educational and crosscultural perspective linking various scientific, social, and artistic fields; to nurture a holistic view of arts, sciences, humanities, and healing; and to publish and distribute literature on the relationship of mind, body, and nature.

The Institute of Noetic Sciences is a twenty-five-year-old research, education, and membership organization that specializes in the research of consciousness. This book is a product of the Institute's Causality Program.

Library of Congress Cataloging-in-Publication Data

Harman, Willis W.
    Biology revisioned / Willis W. Harman, Elisabet Sahtouris.
       p.   cm.
    Includes bibliographical references.
    ISBN 1-55643-267-4 (pbk.)
    1. Biology--Philosophy. 2. Consciousness. I. Sahtouris,
Elisabet. II. Title.
    QH331.H364 1998
    570'.1--dc21   97-46579
                                       CIP

1 2 3 4 5 6 7 8 9 / 00 99 98 97 96

BIOLOGY
REVISIONED

This book is
lovingly dedicated
to the memory of
Willis W. Harman.

*"Perhaps the only limits
to the human mind
are those we believe in."*

—Willis Harman
Global Mind Change
1988

Table of Contents

Foreword

During the last thirty years of his life, I was privileged to know and work with Willis Harman. Throughout this fruitful period, I watched him dive into one after another of the great ideas of our time—as he sought to understand the societal transformation he was so confident we are living through. As a futurist, his interest in this shift moved from a general appreciation of social change to certain specific areas: first he was intrigued by *the power of our belief systems,* and then sequentially he turned his attention to *intuition, creativity, the meaning of work, the changing role of business* . . . and finally, to the roots of one of the most influential forces shaping Western culture, *science.* Each exploration led to new ideas, creative syntheses—and the authorship of stimulating new books sharing what he had learned. Underlying all of these works was a conviction that we need to examine our deepest assumptions if we are to actualize our fullest human potentials.

Around the time of his seventieth birthday, a friend asked Willis what topic he would most enjoy exploring during his remaining years. His choice was vigorous and immediate. He wanted to probe the core assumptions underlying modern science, to see whether these were based on immutable principles, or reflected the historical conditions in which modern science arose some four hundred years ago. His interest was highly pragmatic: many of the topics most deeply relevant to humankind are excluded from study using a science that is based on reductionism, positivism, and materialism.

Willis had developed the background for this new direction during his nearly twenty-year tenure at the Institute of Noetic Sciences. In particular, the seven-year-long "Causality Project" arose from this choice. Under Willis's direction, the project drew together an impressive multi-disciplinary, international group of scientists, philosophers and researchers to probe the fundamental belief system underlying modern science. Together, they explored what science might look like if a different set of rules were chosen. What would happen, they inquired collectively, if we *began* with the assumption of unity rather than separation? The resulting "wholeness science" could include participative methodologies, focus on understanding as well as prediction and control, and be radically empirical in that no area of human experience need be excluded from scientific study.

Biology Revisioned arose from this intellectually rich environment. The life sciences provide a particularly meaningful area of inquiry in which to explore the practical consequences of these new ideas. Willis had become convinced that recent developments in biology, coupled with major new insights and theories from other branches of the sciences (particularly quantum physics, complexity theory, non-linear dynamics, and systems and field theories) were pointing the way towards an expanded science that could include our inner experiences on a par with the physical world. With typical enthusiasm and curiosity, he set out to immerse himself in biology and to discuss his emerging ideas with a most distinguished and thoughtful group of biologists.

To bring these ideas to completion, he sought an open-minded and discerning biologist, Elisabet Sahtouris, to engage in a dialogue centered on the science of life. In keeping with the original meaning of the word *dialogos,* their process reveals an emergent meaning deeper than what either might have come to alone. The biology that Willis and Elisabet have revisioned establishes a theoretical framework to allow for interconnectedness, and as such complements both quantum physics and Eastern spiritual philosophies. Likewise, it includes the interplay of human consciousness and the physical world as necessary to a full understanding of reality and our place in it—a meeting ground that is both researchable and knowable.

The Institute of Noetic Sciences, where Willis served as President for many years, is committed to an on-going inquiry into the nature of reality, with a particular emphasis on the nature and potential of human consciousness. Balancing rigor and open-mindedness, our inquiry respects and utilizes all ways of knowing; including the reasoning processes of the intellect, perception of our experiences through the senses, and the intuitive, spiritual or inner ways of knowing. A primary goal is to expand the boundaries and methodologies of Western science in order to more fully accommodate the full spectrum of human experience. This exciting shift of perspective is reflected in Willis's final book. *Biology Revisioned* is a fitting capstone to his long and inspiring career, spent engaged with some of the most potent ideas of our time: It is offered to the reader in service to the continuing story of life.

—Winston Franklin, President
The Institute of Noetic Sciences

Preface

One hears many murmurs these days that the biological sciences—and hence our understanding of life—may be headed toward fundamental change. David Depew and Bruce Weber (1985) postulate a need for *The New Biology* and the *New Philosophy of Science.* Ernst Mayr (1988) bids us look *Toward a New Philosophy of Biology.* Robert Augros and George Stanciu (1987) write of *The New Biology: Discovering the Wisdom in Nature.* Richard Levins and Richard Lewontin (1985) describe how things appear different to *The Dialectical Biologist.* Henri Bortoft (1996) writes of *The Wholeness of Nature: Goethe's Science of Conscious Participation in Nature.* British biologist Brian Goodwin (1994b) urges us toward "a science of qualities."

While many persons in the biological sciences sense a creative ferment and anticipate some major development, there is less agreement as to what this "new biology" might be. At least three widely differing views can be discerned:

View 1. Quantum physics and complexity theory provide new insights of such power that their implications will lead to a qualitatively "new" biology; no fundamental epistemo-logical modification appears to be required.

View 2. There is need for a more holistic biology, character-ized by recognition of wholes being more than the sum of their parts, by "emergent" qualities not reducible even in principle to the physical sciences, and by a more participatory epistemology.

View 3. Even if View 2 accurately describes the next phase, it is really a bridge to a still more radical biology accommodating the insight that something resembling consciousness appears to be present as a substrate, so to speak, of physical reality.

View 1.
New tools.

In the first of these three interpretations, the chief new empower-ing insights are (a) that intracellular processes are at energy and information levels such that quantum physical concepts are nec-essary to understand them, and (b) that complexity theory reveals that information can be created, giving a new insight into a central puzzle of evolution—namely, the origin of those quali-tative shifts that provide the raw material for natural selection, and that seem to occur with little or no selection.

Complexity theory, in particular, appears to bring a radically new possibility to evolutionary theory. It is widely recognized that the Darwinian hypothesis of the origin of new species—ran-dom mutations followed by natural selection—does not ade-quately fit the observed data. The gradual change and continuity that this implies is simply not found in the fossil record. New types of organisms appear upon the evolutionary scene, persist for varying periods of time, and then become extinct. There is no widely accepted explanation for this relatively sudden emergence of novelty. It seems to require some other explanatory principle than natural selection operating on small variations. The appearance of order out of chaos in complexity theory seems to give a promising lead toward accounting for this phenomenon.

There is multifold evidence for a fundamental self-organizing force in living systems, from the smallest to the largest conceivable organisms, which remains unexplained by physical principles. Living systems exhibit a tendency toward self-organization (e.g., homeostasis, intricate patterns in flowers, butterfly wings, etc.); toward preservation of integrity (e.g., healing and regeneration, ontogenesis from a single fertilized egg to an adult organism); toward survival of the organism and the species (e.g., complex instinctual patterns for protection and reproduction). Attempts to explain this self-organizing force in terms of genetic programs have been unconvincing. However, again the demonstration, in complexity theory, that order can arise out of seeming chaos, offers new hope of understanding self-organization in terms of epigenesis.

The cumulative effect, over time, of this self-organizing tendency in evolution shows up as apparent purposefulness, or as teleological in nature. Teleology has been an unacceptable concept to most biologists; nevertheless, it strains the imagination to conceive of a picture of evolution that does not include at least some sort of survival instinct. Using models from complexity theory may relieve some of the discomfort with this teleological appearance.

Among mainstream biological scientists, this first view would undoubtedly be the favored interpretation—insofar as they would accept any concept at all of a "new biology." It builds on the demonstrated efficacy of the scientific paradigm in its present form, and does not open the door to "soft" speculation. However, a number of biologists have recently opined that the "new biology" must go much further and must in fact amount to a definite shift in perspective rather than simply the addition of new tools to reductionistic science.

View 2.
A more holistic biology.

The second interpretation of the "new biology" moves much further in the direction of holism. It includes, first of all, a reinstatement of the organism. In this view, the organism is not adequately explained in terms of genes and their products; epigenetic factors need to be taken into account in the development of the organism. Organisms are integrated living systems, not just complex machines controlled by the genes carried within them. An organism is a functional and structural unity in which the parts exist *for* as well as *by means of* one another in the expression of a particular uniqueness. Each part, in other words, is dependent upon other parts, and serves other parts as well as the whole. The parts are not made independently and then assembled, but arise as a result of interactions within the developing organism, and between the organism and its environment. Just as physics has different organizing theories for microscopic particles (quantum mechanics), macroscopic fluids (hydrodynamics), and stars and galaxies (relativity theory), so biology may be best served by molecular biology augmented by a theory of organisms as distinctive entities in their own right, with characteristic types of dynamic order and organization (further integrated into co-evolving organism-environment systems, and so on).

The nature of an organism is usually described in terms of the species to which it belongs. Among the most distinguishing qualities of a species are its spatial pattern (its form, or morphology) and its characteristic patterns of activity (feeding, mating, migrational, etc.). The study of biological form, in space and time, is the beginning of what Brian Goodwin has termed a "science of qualities" that complements and extends the powerful science of quantities (in particular, molecular biology), which tends to dominate the field at present.

In this more holistic biology, the organism is not separable from its environment. Gail Fleishaker has proposed the term "ecological individuals" to differentiate this concept from the isolable organism. Life appears and persists not as a sum of multiple discrete entities but as a single ecology—the multileveled embedded consequence of life's active and continuous metabolic operation. It is the operational unity of these overlapping systems that marks living systems spatially and temporally as ecological individuals. Spatial unity of operation is evidenced in the expansion of cells, populations, and communities into ecological space, using the waste of others as food. Temporal unity of operation is evidenced in the extension of intimate associations retained among living systems over time. Symbiosis and the endosymbiotic origin of nucleated (eukaryotic) cells are examples of the unitary and integrated operation of partners.

Another aspect of this more holistic biology is referred to with such terms as "participative" or "qualitative" methodology. It is sometimes spoken of in terms of "first-person" science, as contrasted with ordinary, "third-person" science. This is basically another way of focusing on the scientist's personal experience of the object and recognizing that this way of perceiving creates its own form of intimate knowledge, which can be shared only through something like apprenticeship—certainly not through mathematical equations and scientific prose. Such a science had been urged by Goethe and later by Rudolf Steiner. Brian Goodwin, in particular, has written about it extensively. Something like it can also be found in the "science" of indigenous peoples. It is in these places that we can look for clues to the kind of understanding to be provided by a new "science of qualities."

One central concept of holistic biology is that there is a natural hierarchy of molecules→organelles→cells→tissues→organs→ organisms→societies. Arthur Koestler introduced the term "holons" in "holarchy" to avoid some of the negative connotations of the word hierarchy, and we adopt this term throughout this book. Each biological, while unified by a central logic of living entities, displays its own complex dynamics and emergent properties—qualitatively different from anything found among its components. At the same time, these hierarchical metaphors do not contradict the basic unity. In the holarchy, chaotic behavior at one level can give rise to distinctive order at the next level. This becomes a key factor in the understanding of morphology, behavior and culture. (As H. Patee asserted, "Hierarchical control is the essential and distinguishing characteristic of life.)

One of the distinguishing characteristics of a holistic biology is in the metaphors used. The reductionistic neo-Darwinism literature is replete with such metaphors as "information" in the DNA, genetic "program," competitive interactions among species, survival of the fittest, and "selfish genes" survival strategies. In this view of evolution, species either work or they don't; they have no intrinsic value or holistic qualities, and the metaphors reveal this. In a holistic biology, by contrast, we find such metaphors as continuum, cooperation, altruism, play, creativity, agency, and "life at the edge of chaos." Apparent purposefulness is not necessarily something epiphenomenal, to be "explained" in terms of chemistry and physics, but an observable quality to be included in theories at the higher hierarchical (hol-archical) levels.

The holistic view recognizes that in morphogenesis, or the creation of form, the emergent order is generated by distinctive types of dynamic processes in which genes play a significant but limited role. Control of the process of morphogenesis includes epigenetic factors and processes that have yet to be understood. An important part of the new science is study of the dynamics of emergent processes. The central quality of the evolutionary process is creative emergence.

In this view, even microörganisms—as shown by the laboratory work of John Cairns and the evolutionary theorizing of Lynn Margulis—appear to be problem-solving organisms, in spite of the obvious fact that they have no brains or central nervous systems.

Probably the issue that most divides this interpretation of the "new biology" from View 1 has to do with the observed tendency of all organisms to self-organize *(autopoiesis)*. In the former view, self-organization is epiphenomenal, to be explained, most recently, through complexity theory. In the holistic view, self-organization is an emergent quality to be studied in its own right, not reduced to something lesser.

View 3.
New epistemological and ontological assumptions.

The third view, which is what emerges in this book, is distinguished from the other two by its suggestion that the self-forming characteristic of living beings, as well as numerous other biological puzzles, require a re-examination of the metaphysical assumptions that tend to underlie all of Western science.

This third interpretation of the "new biology" is related to the long-standing dispute in philosophy between realism and idealism. The French philosopher Henri Bergson attempted to resolve this issue with his philosophy of process. In particular, Bergson insisted on dealing with two basic modes of knowing— that from physical sense data and that from deep intuition— rather than relying solely on the former. Or as Owen Barfield put it, "There is indeed only one world, though with both an inside and an outside to it, only one world experienced by our senses from without, and by our consciousness from within."

Various scholars, from George Wald to Ken Wilber, have suggested that the biological sciences would take a rather different form if one assumed the possible presence of something like consciousness in accounting for phenomena. In particular, such an assumption would strongly affect interpretations of evolution, adaptation, morphogenesis/ontogeny, immune system functioning, instinctual behavior patterns, and organisms' ability to distinguish "self" from "not-self." Lynn Margulis has hinted at this assumption in her discussions of the "clever" adaptations and innovations of microörganisms in the early history of life on the planet. The overclaims of influence of the DNA seem related to omission of anything like consciousness from our understanding of organisms and inheritance. The long-standing antipathy toward concepts of teleology and purpose, and wariness about the concept of intentionality, seem also to be related to this omission. Indeed, all areas of biological science would be affected by admission of such an assumption regarding consciousness— including the question of the origin of life, the appearance of novelty in evolution, self-assembly and organizing systems, the adequacy of the neurosciences in their present form, and the explanatory power of biochemistry and molecular biology.

In this book we approach this matter from the standpoint that it is an historical accident that physics should have come to be the generally accepted root discipline within science. To a greater or lesser extent, the other sciences attempt to imitate physics, and the assumption is widespread that ultimate explanations are in terms of fundamental particles and basic fields.

But what if biology rather than physics should be taken to be the root discipline? If science were to start with biology, dealing with wholes prior to parts would seem the natural thing. Holistic concepts such as organism and ecological systems would be the starting points, rather than discrete "fundamental particles." It would not be a matter of great surprise to find, through the revelations of quantum physics, that everything is connected; one never would have assumed the separation in the first place. Since our inner mental experiences comprise our most direct contact with the greater reality, information from the physical senses being in a sense secondary and indirect, consciousness would naturally be of central interest in biological study. There would be no reason to assume it epiphenomenal or to be explained in terms of physical functions.

Although Arthur Koestler introduced the terms "holon" and "holarchy" many years ago, attempting to deal with these holistic aspects, the terms have not exactly become part of the common parlance. However, Ken Wilber's recommended ontological stance (which fits with a wide range of human experience) is to consider reality as composed of "holons"—each of which is a whole and simultaneously a part of some other whole (1996). One of the characteristics of a holon in any domain is its *agency,* or its capacity to maintain its own wholeness in the face of environmental pressures that would otherwise obliterate it. It has simultaneously to fit in as a part of something else, with its *communions* as part of other wholes. Its further capacities are *self-dissolution* and *self-transcendence.* A holon can break up into other holons. But every holon also has the tendency to come together with others in the emergence of creative and novel holons; evolution is a profoundly self-transcending process. The self-transcending drive produces life out of matter, and consciousness out of life.

This new ontological stance of "holarchy" (holons within holons within holons . . .) has many attractive aspects. For example, it appears to successfully resolve many of the time-honored puzzles of Western philosophy (the mind/body problem, for example, and free will versus determinism). Since everything is part of the one holarchy, if consciousness is found anywhere (such as at the level of the scientist-holon), it is by that fact characteristic of the whole. We can neither rule it out at the level of the microorganism, nor at the level of the Earth, or Gaia.

Bringing the ontological stance into the biological sciences turns out to involve some fascinating and surprising implications. The biological sciences contain within them a wide assortment of enigmas. Living systems, from the smallest microbes to the largest organisms, exhibit self-organization; all of life is basically defined by this self-generating, self-maintaining criterion. Although autopoiesis applies all the way from the most simple single-cell life form to Gaia, it has never been explained by physical principles. Within the context of holarchy, there seems much more promise of it becoming understandable.

Or consider the puzzle of ontogeny, the creation of form in the development of an organism. Brian Goodwin asserts that contemporary biological explanations tend to describe only necessary conditions. Genes cannot account fully for the morphogenesis of organisms; DNA is a necessary but not sufficient determiner of form. A more complete explanation involves three potential levels of explanation, all complementary, no one contradicting another:

molecular (genetic information), structural (dynamic principles of form in living organisms), and creative emergence (implying something like a substratum of consciousness). The first of these is well accepted, and considered by many scientists to be sufficient. The second is less universally appreciated, but would generally be considered to be a legitimate area for exploration. The third is typically considered to be vitalistic nonsense, long discarded by the mainstream scientific community; however, in the holarchic context it appears to be an essential factor.

There are many further examples. When scientists, such as Lynn Margulis and Mae-Wan Ho, began watching live modern bacteria, not to mention larger organisms, and saw them responding to changes in the environment by appropriate mutations in very short order, it seemed impossible to deny intelligence and intentionality. Consider the puzzle of recognition; it's a little difficult to see how that ability can be explained through a "program" in the DNA. Many other forms of innate behavior patterns (what used to be termed "instinctive behavior") amount to fundamental mysteries as well. It turns out that when the data relating to evolution are thoroughly examined, neo-Darwinism can be challenged as a significantly misleading way of seeing nature.

By conceptualizing and modeling the entire universe as a holarchy containing smaller holons in continual co-creation, its intelligent and conscious evolution makes far more sense than within the old mechanical model of a nonliving universe. In this view, consciousness is not considered to be an emergent property of evolution. Within this view there is no occasion to raise the question how consciousness could emerge from nonconsciousness, any more than life from nonlife.

We moderns are reluctant to recognize Western science as an artifact of European culture, rather than the unique and best road to truth. But we are quite aware that different perspectives are found in Eastern and indigenous cultures. There really is no valid reason to suppose that reductionistic science by and of itself can ever provide an adequate understanding of the whole.

This book's aims.

This book is an honest accounting of the intellectual adventures of the two authors as we kept raising fundamental questions. It incorporates insight we received from a number of our friends. It deals primarily with the question of what kind of biological science will serve us all best, and ends with some discussion of implications of a different worldview for our approach to societal and global problems and dilemmas.

We have chosen to write this book in a semi-conversational mode, because the informality of that mode allows us the psychological freedom to explore some matters about which we might be more cautious if our exposition were to be more formal. Also, by choosing this type of presentation we have attempted to convey some of the excitement we felt as we explored these topics together. We hope that you "catch" it too.

—Willis W. Harman
—Elisabet Sahtouris

Acknowledgements

This book exists because of an idea. But equally, it exists through the efforts of the many minds, hearts and hands which helped bring this idea into its present form. *Biology Revisioned* has benefitted from a long series of conversations that elicited the broad issues to be addressed, pointed out fruitful topics for in-depth exploration, and helped identify inconsistencies of interpretation and alternatives to developing ideas. Among the key contributors were members of the Institute's Causality Project, including George Wald, Eugene Taylor, Roger Sperry, Arthur Zajonc, Michael Scriven, Lynn Nelson, Ilja Maso, Charles Laughlin, Robert Jahn, Brenda Dunne, Beverly Rubik, Harry Rubin, Richard Dixey, Mae-Wan Ho, Brian Goodwin, Richard Strohman, and Vine Deloria, Jr. In particular, the extended circle of colleagues who attended the symposium "Consciousness in the Biological Sciences: Implications for the Biological Sciences of Recent Developments in Consciousness Research" provided food for thought. Participants at the meeting (Ben Lomond, California, 1995) included, in addition to some of those named above, Peter

Corning, Christian de Quincey, Daniel Eskinazi, Bernard Grad, Gary Heseltine, Bruce Kirchoff, Kenneth Klivington, Nola Lewis, Andy Parfitt, Bruce Pomeranz, Glen Rein, Marius Robinson, Peter Russell, Rolf Sattler, Marilyn Schlitz, Linda Shepherd, Dennis Todd, Alan Trachtenberg, and Jan Walleczek. All shared generously of their expertise and information, and continued those discussions long after the meeting ended.

Elisabet Sahtouris's important participation is acknowledged as co-author; her curiosity and intellectual vigor provided an important stimulus as the dialogue and project unfolded. Multiple drafts of the manuscript were exchanged over the Internet and deadlines met, despite distances that were often global.

Thanks to Richard Grossinger, Susan Bumps, Emily Weinert, and Nancy Koerner at North Atlantic Books for their ability to see the relevance of these ideas in their roughest form, and for their comforting expertise in seeing the manuscript through the myriad steps needed to make it into a book, and to Michael Schaeffer for his editing skills. Adrienne Smucker's imaginative talent translated vague instructions—"a picture goes here"—into clear illustrations of key ideas. Stephen Lieper and Jeanette Mathews helped with all the proofreading, corrections, and attention to detail that is an integral part of producing a finished book.

We are especially indebted to Laurance Rockefeller, The Fetzer Institute, Theodore Mallon, Bob Schwarz, David Moline, Marius Robinson, Lifebridge, and the Seven Springs Foundation for their initiatory support of the Causality Project, and thus of this work. Their visionary foresight discerned the value of this unique exploration, and afforded Willis the opportunity to bring it to completion during his last years of life. It is a gift which we believe will bear fruit in the years to come.

ACKNOWLEDGEMENTS

As Director of Research, Marilyn Schlitz's ideas and comments have consistently proven invaluable in shaping the Causality Project as well as this book. And last, a special word of thanks to Nola Lewis, whom Willis entrusted with the task of bringing *Biology Revisioned* into the world for others to share when he learned that his illness would not allow him to shepherd it himself.

CHAPTER
ONE

The New Context: From Mechanism to Organism

History shows that the human mind, fed by constant accessions of knowledge, periodically grows too large for its theoretical coverings, and bursts them asunder to appear in new habiliments, as the feeding and growing grub, at intervals, casts its too narrow skin and assumes another, itself but temporary. . . . A skin of some dimensions was cast in the 16th century, and another toward the end of the 18th, while, within the past fifty years, the extraordinary growth of every department of physical sciences has spread among us mental good of so nutritious and stimulating a character that a new ecdysis seems imminent.

—T. H. Huxley (1863)

HARMAN:

We in modern society are seeing signs of a fundamental change in the way we understand ourselves and the way we relate to the universe. This is showing up both within science and in the culture at large. A good place to examine this appears to be in the biological sciences. Biologists have been extremely successful in employing the reductionist, prediction-and-control focused paradigm of Western science in molecular biology and biotechnology. They have, through the neo-Darwinian concepts of evolution, had a strong influence on the entire modern "story" of our world. Yet I'm struck by the fact that, of the seven major questions addressed by the biological sciences, a number of the accepted explanations are now open to serious question.

A science is defined by the questions it asks. Among the many questions and issues with which the biological sciences deal, a handful stand out as particularly basic to the inquiry. These include:

1. What accounts for the *emergence of life?* How did living organisms arise out of nonliving material?

2. How do organisms *reproduce* (and pass on characteristics to their offspring)?

3. How have the higher forms of life *evolved?* What accounts for novelty in the evolutionary process?

4. How do organisms *develop?* What is the explanation of form (structure and behavior)?

5. How do organisms *recognize* themselves as opposed to others? (And how do they recognize others, such as potential mates of the same species or members of the same colony?)

6. How do they *feel* (the problem of consciousness awareness)? When, where, and how in the evolutionary process does consciousness appear?

7. How can we explain the *appearance of purpose* (teleology) throughout the biological world?

The answers to some of these, at least, appear to be in process of revision at the present time. Let's start with the emergence of life. The spontaneous appearance of certain organic chemicals when laboratory conditions are made to simulate conditions on Earth in the prelife period has, over the past half century, prompted hope that scientists might be only a few steps from the creation of life. Yet the gulf between nonliving and living remains unspanned and shrouded in mystery. Doubt increases that attempts to explain the origin of life solely on the basis of chance creations of complex molecules will succeed.

The problem of reproduction is generally considered to be largely solved, primarily through the conceptual breakthroughs of Gregor Mendel and the team of Watson and Crick. Around 1920–30, under the stimulus of Mendelism and of the developing science of biochemistry, biologists began to realize that many of the properties of growing and adult organisms are partly due to the presence, both in the original fertilized egg (zygote) and in every cell that arises from it, of minute chemical units possessing a highly specific structure (RNA, DNA). These carry the hereditary information; they divide and replicate when cells divide in the process of creating the new adult individual; they convey the information in the gametes (sperm and egg cell) when these unite to form the zygote. Yet a fundamental question remains. The phenotype (the total characterization of the adult organism) is generally assumed to be determined by the genotype (the actual genetic makeup) and to be significantly influenced by the environment within which the organism develops. But is that sufficient to explain, for example, complex instinctual patterns for protection and reproduction?

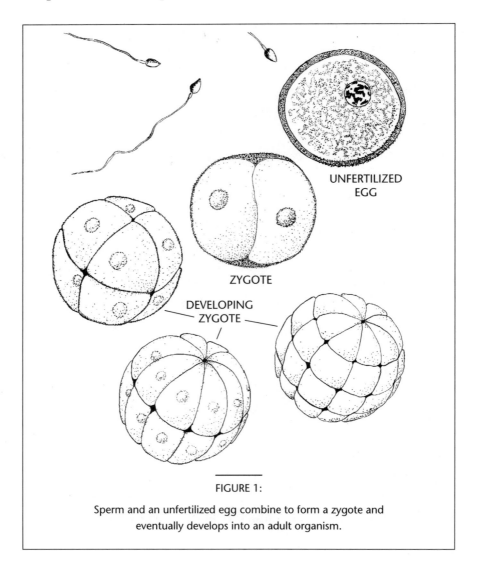

UNFERTILIZED EGG

ZYGOTE

DEVELOPING ZYGOTE

FIGURE 1:

Sperm and an unfertilized egg combine to form a zygote and eventually develops into an adult organism.

The mechanisms of evolution are also assumed to be largely understood, though both local and fundamental mysteries remain, as I'd like to explore later.

A great deal is known about the mechanisms involved in morphogenesis, the process by which the zygote transforms

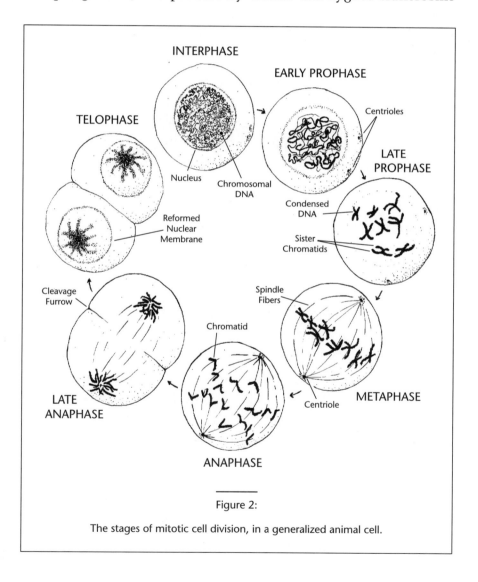

Figure 2:

The stages of mitotic cell division, in a generalized animal cell.

through a well-defined sequence of stages into the adult form. It seems rather well established that this process can in no way be directly controlled solely by some sort of code in the DNA. As the original one-celled zygote repeatedly divides and the embryo begins to form, the individual cells become subject to morpho-regulatory molecules that regulate adhesion and movement. Starting out from a small ball of identical cells, the descendants of these cells move around and gather together in various ways, and eventually become differentiated into things such as liver cells and muscle cells and neurons. Most theories of embryonic development in the last seventy years have attempted to make something of the notion of positional information—the idea that the present location of a cell and its present activity provide much of the information regarding what it is to do next. Metaphors of "field" and "gradient" and "spatial waves" of chemical concentration have dominated embryology. Yet in spite of much being known about detailed mechanisms, the ordering principle behind the mechanisms is still unknown. The puzzle of morphogenesis remains one of the key enigmas of biology.

Recognition, at the molecular and cellular levels, is understood in terms of the formation of antibodies. Which is to say, even at the cellular level I know who I am, and can recognize an alien cell that is "not me." The body's immune system can recognize tissue as being from itself or some other organism. White blood cells do an amazing job of recognizing invaders of various types. That's pretty remarkable, but knowing about the mechanism (antibodies) does not resolve the entire mystery. How organisms recognize the opposite sex of their own species at mating time is another mystery. Pheromones—scent attractors—play a role, but the explanatory power of this mechanism hardly explains the observable phenomena.

As for consciousness, this mystery appears to be in some way at the heart of all the others. And intentionality, the appearance of purpose, is perhaps best thought of as part of the same puzzle. It is becoming increasingly apparent that an adequate dealing with the issue of consciousness is not just something scientists will get around to in time; it is a basic challenge to the completeness of Western science that must be dealt with. Somehow there will have to be a more satisfactory accommodation of "consciousness as a causal reality" (to use Nobel laureate Roger Sperry's apt phrase); this is certain to have important implications for the biological sciences. All this suggests that revolutionary changes will be happening within the field of biology and that those changes are bound to have profound implications for society.

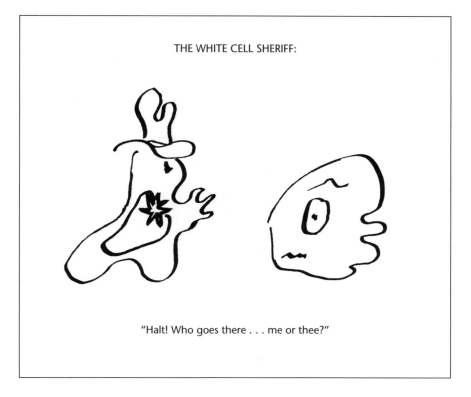

THE WHITE CELL SHERIFF:

"Halt! Who goes there . . . me or thee?"

SAHTOURIS:

I believe they are indeed happening. Biology is very much a part of the whole shift of emphasis in science from mechanics to organics. More than one observer has pointed out that biology and physics have been on two trains running on parallel tracks but in opposite directions—physics moving to a quantum field worldview in which the cosmos is seamless, fluid and self-organizing, with consciousness integral, primary and causative, while biology has moved with equal persistence along the mechanistic reductionistic track laid down by Newton, now focused in microbiology.

But all this is changing rapidly. Our growing understanding of living systems is really pushing us to reconceptualize biology accordingly. We have to ask anew how to envision the whole process of life, of living systems, and how to proceed most fruitfully in our study of them. And as you say, these revolutionary developments are bound to have a profound impact on society.

Of course, science has always influenced society profoundly; one might even venture to say that it has done so more than any other human endeavor, since it has altered all of them. But our new insight is that scientists in general and biologists in particular, with all their theories and experimental findings, exist as an influential component of the larger living system of all human culture, which in turn is embedded within the vast and complex living system of our planet. Biologists, indeed all scientists, live in a participatory universe. This represents a sharp departure from the old notion of the scientist as an objective—that is, distant—value-free observer and describer of "how things are."

To acknowledge the role of biological sciences as influential in the changing dynamics of human culture is exciting to me. The emerging biology could actually provide the basis for reorganizing human society more holistically, and for solving our critical world problems using information accumulated by our planet over five billion years of evolutionary experience. It can, and I believe it must. As Roger Sperry has said, "In the context of today's worsening global situation and our imperiled future, perhaps the most important feature of the described new outlook of science is its provision of a prescription for long-term, high-quality survival and a way out of our current global predicament."

The new sciences of complexity and chaos have taught us to look for overall pattern rather than only for the reductionist specifics. We have so avidly pursued the latter that we have lost the forest for its trees—or rather for its leaf stomata and root hairs. To really get the picture of this all-important transition in biology and other sciences, I would like to consider the context in which science developed, for context is a guide to understanding phenomena; context gives us explanation and meaning.

Consider science as a human activity arising from such basic questions as "Who are we?" "Where do we come from?" "Whither do we go?" Our need to give ourselves a frame of reference, to orient ourselves in our world, our cosmos, and to guide our actions within it are as natural and as necessary to our life as our birth and our breathing. In our need to order the complex unceasing perceptions that seem to come from both within and outside ourselves, to give life order and meaning, we create human worldviews and cultures comprising religion, science, economics, politics, art, and ethics with all their concepts, rules, and fashions over time and place. Our worldviews inspire us to action, determine our security, and guide our flexibility, both as individuals and as cultures. We defend them against each other, sometimes unto death; we are also capable of changing them in response to new situations or new information.

As things stand, our main metaphors for life and the universe at large come from our invented mechanics. We discuss the machinery of nature and extend the metaphor beyond science to our socioeconomic systems, which we hope to run as "well-oiled machines."

In fact, human welfare and progress are measured in terms of the technological transformation of the natural world to human use; science is funded essentially to assist in this process. All this in turn has influenced the kinds of research that scientists can get funded, and the kinds of results they obtain. Yet it has not been easy for scientists to see the restrictions thus imposed upon their cherished freedom.

Where did our scientific worldview come from? Historically, science was coupled from its inception with this concept of controlling the natural world through machines. In this heady time, Descartes set the scientific worldview as one in which God was not only a mathematician, as in Plato's and Galileo's views, but also a Grand Engineer, inventing all the machinery of nature and giving man—apparently his favorite robot—a piece of godmind so that he, too, could invent machinery. However strange this seems to most of us now, it was a logically satisfying view of things, and has persisted to the present in the worldview of some well-known thinkers, such as Buckminster Fuller, and biologists. By dropping God from its cosmology as an unnecessary hypothesis, scientists changed this worldview into one that is, in fact, logically incomplete. Western science took a leap into illogic by assuming that the machinery of nature could form itself without an inventor, by accident, in a purposeless world within an ultimately doomed universe. Not surprisingly, this scientific view of a meaningless universe has had obvious social consequences, such as the existentialist movement and the present-day rise of fundamentalism, with its objection to scientific accounts of evolution.

From the perspective of any other age or culture, or of such contemporary groups as the fundamentalists or creationists, a cosmology in which all living things are assumed to be assembled by accident from fundamentally nonliving parts, to function as marvelously complex machinery, is neither adequate nor satisfying. It is a worldview that violates even our own dictionary definitions of machinery as intentional, purposive constructions, and flies in the face of the obvious: that machinery does not grow on trees. It also violates the calculations of probability theory coupled with our estimates of the total amount of universal matter and the time it has existed. Thus, even in scientific circles, we are finding this an increasingly strange and unsatisfying creation story. In fact, I think we are rapidly coming to wonder that it was taught in our institutions of higher learning at all.

It is interesting to note here that the concept of a "nonliving" universe was invented for the dualistic worldview of Western science by Descartes and has never, to my knowledge, appeared in any other historic or contemporary culture. It seems to have made sense in the heady excitement about nonliving mechanisms being invented at that time—an infatuation with our own devices that made us project them onto the whole universe.

It reminds me how, as graduate students, we were taught that "anthropomorphism"—projecting human characteristics onto living organisms, especially other animals—was a heresy, while at the same time we were required to learn and profess what, though it was not named at that time, I came to call "mechanomorphism" (a term I later found was also used by historian Morris Berman)—the projection of mechanical characteristics onto animals and all nature. Mechanics is an invention of just one species, man or *anthropos,* I reasoned, and so the mechanomorphism required of us was really a kind of secondhand anthropomorphism and was therefore even worse! How could we humans project the characteristics of our anthropic inventions onto all nature, all the universe? Was nature at large not likely to be more like us naturally evolved creatures than like our machines?

Biology is the study of life, which in the Western scientific worldview means the living organisms on the surface of the Earth. The Earth itself is held to be nonliving matter, studied by geologists, chemists, and physicists. The tasks of biology, which now includes such subsciences as genetics, physiology, microbiology and ecology, have been to describe (simplifying a bit the list you gave earlier) (1) how life originated from nonlife, (2) how it evolved and speciated over time, (3) how individuals develop and function, and (4) how species interrelate. The methods of biology have been largely adapted from those of the other sciences: isolating parts and functions as though investigating machinery.

With the current shift from mechanics to organics, biological inquiry is being put into a different light. I believe that the new biology will take on a very exciting new task: to understand how the living system that is our rapidly globalizing human culture has evolved historically within a planetary living system, how it functions, and how to understand its crises biologically. This is what Jonas and Jonathan Salk (1981) attempted and hoped to promote among other biologists.

All of us agree that our world is now in crisis—ecologically, economically, demographically, politically, spiritually, culturally—everywhere one looks. For many, this is overwhelming; we simply cannot see our way out of so many problems coming from so many directions. But do they really stem from different sources and just happen to coincide? Or do they share a common factor or underlying unity that can help us make sense of them all?

Consider that all of these problems exist in living systems, and that the Western scientific worldview shaping world culture has not been oriented to the understanding of living systems in their own right. Rather, it has regarded them as mechanical systems, essentially like the invented technological devices of our industrial and post-industrial world. Even the "laws of nature" discovered by science have found their confirmation more in technological prowess than in the prosperous and happy survival of our species.

From my perspective as a biophilosopher, our scientific vision of both nature and human institutions as mechanisms, and our efforts to "engineer" human society into this mold as political and economic machinery, has blinded us to the nature of the living system we are. As a result, we have a marvelous science for developing technology, but not yet a good science for developing a healthy, ecologically sound and sustainable human living system.

Why was it that the biological sciences were relegated to second-class citizenship within the reductionistic mechanistic worldview of classical physics? Was this due to an accident of history that shaped all of science? What might have happened if Galileo had looked down through the new lenses into a microscope, at living things, instead of looking through his telescope at the sky, which since ancient Greece had been seen as celestial mechanics? Perhaps biology would have become the leading science, and physics would have had to fit itself into a concept or model of a living universe with laws of living nature. Perhaps we would then have today a far better understanding of the nonmechanical nature of living systems. But in Galileo's time, the understanding of living nature was losing ground due to its association with women accused of witchcraft, and men were beginning to focus on technology, in which physics was far more useful.

Historically, then, biology had to fit itself as a lesser science into the paradigms of physics. (Consider, for example, the notion of "reverse entropy," which had to be coined to make biological systems fit into an overall physical model.) Only now, when our world is increasingly seen as an endangered living system, can we hope that biology will stand on its own and gain its much needed influence.

HARMAN:

That's a most intriguing idea, that biology rather than physics should be the root science. If science were to start with biology, dealing with wholes prior to parts would seem the natural thing. Holistic concepts like organism and ecological system would be the starting point, rather than discrete "fundamental particles." We would not find it a matter of great surprise to find that, through arcane revelations of quantum physics, everything is connected; we never would have assumed the separation in the first place.

Also, humans are living organisms, and there's no sharp demarcation between us and other organisms. There is no reason to doubt that the kind of mental functions we find in ourselves are similar to mental functions in other animals—as any pet owner would testify. Perhaps this extends even to plants. There is an amazing amount of evidence (almost totally neglected by academic scientists) to suggest that plants have intent and participate in psychic and emotional relations with their environments—including humans (Peter Tompkins and Christopher Bird, 1973).

Since our inner mental experiences comprise our most direct contact with the greater reality, information from the physical senses being in a sense secondary and indirect, consciousness would naturally be of central interest in biological study. There would be no reason to assume it epiphenomenal, to be explained in terms of physical functions. This has fascinating ramifications.

SAHTOURIS:

It surely does! We have been taught that our planet and all its creatures were not conscious or intelligent before humans appeared. We have been taught to look at nature as though it really were assembled from particles into parts and from parts into wholes, like machinery. We even speak of assembling a global brain out of individual human brains, as though these brains of ours were actual assemblies of neurons put together like computer chips. Doesn't it make more sense to think of them as the dominant part of gradually evolved nervous systems that began with extremely simple neuronal loops and were more complete at every stage of evolution up to our own, always in ontogenic derivation from a single cell? It is time to put such concepts into new perspective, with the shift of emphasis from mechanics to organics, from reductionism to an expanded holistic view of living systems, including Earth as Gaia (James Lovelock, 1979) and post-Darwinian models of intelligent evolution.

HARMAN:

The basic dualistic trap in which Western science has been stuck, as Ken Wilber in particular has emphasized, is to consider the subject doing the mapping as separate from the map. Taking off from physics, as Western science did, it was possible to accomplish a great deal without having to worry too much about the dualistic trap. However, if one starts with biology, it's another story; the trap is there in front of you from the outset. Getting a more accurate (more holistic, more in terms of "systems") map will not solve the problem. Rather, we must realize that thoughts are not merely a reflection on reality, but are also a movement of that reality itself. The mapmaker, the self, the thinking and knowing subject, is actually a product and a performance of that which it seeks to know and represent. The new science we need would both transcend and include the science we have.

SAHTOURIS:

Emphatically, yes! And that concept of transcendence is going to be very important to our discussion of how intelligence and consciousness can be seen as fundamental aspects of all living entities or systems. I would like to propose that we go back to even more basic concepts than organism and ecosystem to find our starting point for the new biology.

However, first let's acknowledge that the emerging models, methodologies, and perspectives in biology shed entirely new light on the fundamental questions of biology that we spoke of earlier. The essential features I see coming into this new biology are:

1. *autopoiesis,* or self-creation, as the fundamental definition of life;

2. the understanding of living systems as embedded within other living systems, all the way from microcosm to macrocosm—call them holons within holarchies, to use the elegant terminology of Arthur Koestler;

3. the acceptance of a multilevel reality, including far more than our four-dimensional world—a reality in which consciousness and morphic fields, for example, are not only legitimate areas of investigation but potentially fundamental aspects of biology's epistemological and ontological framework; and

4. the recognition of intelligent processes operating in evolution (phylogeny) and in the development of the individual organism (ontogeny).

Obviously this represents a very fundamental revisioning, although biological knowledge gained within the old framework will largely remain as valid as did physical knowledge gained within the Newtonian framework after Einstein and others expanded the domain of physics to relativity and quantum fields.

I hope that this discussion will enable us to bring into clearer focus the rapid conceptual and practical evolution biology is now undergoing.

HARMAN:

If you are correct that this "fundamental revisioning" is taking place (or is about to), this really amounts to a "holistic revolution" in the biological sciences.

But you have laid out a lot to explore; we may have to take it a bite at a time. Let's start with the holons. These words "holon" and "holarchy" have not exactly become part of the common parlance, although Arthur Koestler introduced them many years ago; you incorporated them in your work (1989), and Ken Wilber adopted them for his explanation of "how things are" (1995).

You and Wilber recommended the ontological stance (which fits with a wide range of human experience) of considering reality

as composed of "holons," each of which is a whole and simultaneously a part of some other whole. There is nothing that isn't a
holon. (For example, atom→molecule→organelle→cell→tissue→
organ→organism→species→ecosystem→Earth→galaxy are each
holons.) According to Wilber, one of the characteristics of a holon
in any domain is its *agency*, or its capacity to maintain its own
wholeness in the face of environmental pressures which would
otherwise obliterate it. It has to fit in simultaneously as a part of
something else, with its *communions* as part of other wholes. These
are the holon's "horizontal" capacities.

Its "vertical" capacities are *self-transcendence* and *self-dissolution*.
A holon can break up into other holons and in some sense cease to
exist. But every holon also has the tendency to come together with
others in the emergence of creative and novel holons. Evolution is
a profoundly self-transcending process; it has the utterly amazing
capacity to go beyond what went before. The drive to self-transcendence appears to be built into the very fabric of the universe. The
self-transcending drive produces life out of matter, and mind out
of life.

Holons relate "holarchically." (This term seems advisable
because "hierarchy" has a bad name, mainly because people con-
fuse natural hierarchy [inescapable] with dominator hierarchy
[pathological].) In general, for any particular holon, *functions and
purposes* come from levels farther out in the holarchy; *capabilities*
depend upon the next level in. There is no unique holarchy for a
given holon; for one purpose one may want to speak of organisms
within species, and for another purpose, of organisms within
communities. Some holons may not seem to have a function with
respect to the next holon out—parasites in a larger organism, for
instance—yet may serve a function in some holon still further
out, such as the ecosystem.

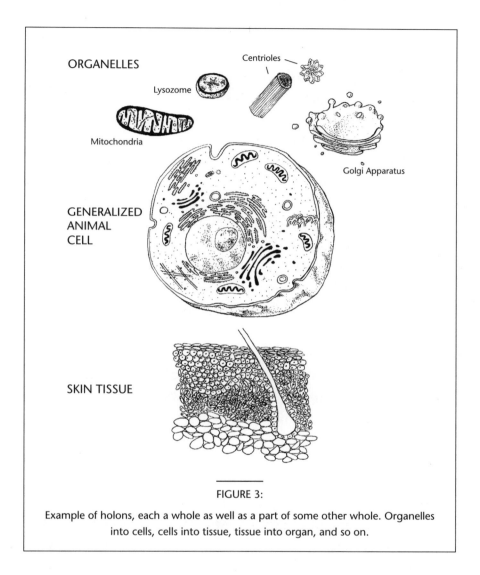

FIGURE 3:

Example of holons, each a whole as well as a part of some other whole. Organelles into cells, cells into tissue, tissue into organ, and so on.

In this sort of picture of reality, the scientist-holon seeking to understand the universe is in an intermediate position. Looking inward in the holarchy (or to the same level, in the social sciences), and exploring in a scientific spirit of inquiry, it is immediately obvious that the appropriate epistemology, or mode of inquiry, is a *participative* one. That is, it recognizes that understanding comes not solely from being detached, objective, analytical, and coldly clinical, but also from cooperating with or identifying with the observed and experiencing it subjectively. This implies a real partnership between the researcher and the phenomenon, individual, or culture being researched—an attitude of "exploring together" and sharing understandings.

Looking outward in the holarchy, it seems clear that the appropriate epistemology involves a holistic view in which the parts are understood through the whole. This epistemology will, for example, recognize the importance of subjective and cultural meanings in all human experience—including experiences, such as some religious or interpersonal experiences, that seem particularly rich in meaning even though they may be ineffable. In a holistic view, such meaningful experiences will not be explained away by reducing them to combinations of simpler experiences or to physiological or biochemical events. Rather, in a holistic approach, the meanings of experiences may be understood by discovering their interconnections with other meaningful experiences.

SAHTOURIS:

This concept of partnership between researcher and phenomenon is very interesting and important. My Native American scientist friends say they must integrate with nature to learn from it, but that is a novel idea to most Western scientists. Still, it happens—I think of Barbara McClintock identifying with her corn, Jonas Salk learning to think like a virus, Lynn Margulis discovering the intelligence and consciousness of her bacteria.

The whole question of self-transcendence in nature, in evolution, is extremely important to the development of our new model, but we need to build the case for such statements as "evolution is a profoundly self-transcending process," and "the drive to self-transcendence is built into the very fabric of the universe." Such statements are clearly inconsistent with neo-Darwinist biology.

Let's continue by first exploring this model of holons in holarchy, which provides us with a foundation to build on. Its power lies in helping us understand interdependence and other forms of interrelatedness both ways: from smaller to larger living systems and from larger to smaller.

I want to emphasize your comment, or Wilber's, that a holon, existing as an entity in its own right, has to fit in simultaneously as a part of something else, "with its communions as part of other wholes." This is very much to the point, and to describe how it works, I adopted the concept of mutual consistency from "bootstrap physics," which tells us that no particle is more fundamental than any other and all, in a sense, create each other. This is a very useful idea for the world of biology as well, since organisms or living systems in holarchies can be shown to determine each other as they co-evolve.

Each holon enjoys relative autonomy (literally, Greek for self-rule) as an individual, but faces the autonomy of the next larger holon as its own 'holonomy' (whole-rule). Thus, each holon looks out for its own interest and integrity, but because the larger holons within which it is embedded do the same, every holon's autonomy is constrained by its holonomy (equivalent to the autonomy of the next larger holons). In this situation, the entire holarchy must necessarily work out a mutual consistency among its holons if it is to survive.

How this endless negotiation of self-interest functions at every level, or in each nested holon, can be seen as the search for dynamic balance or mutual consistency through negotiation between parts and wholes, or rather between holons and holarchies. In the relationship between two people we call a couple, for example, there are three holons: two individuals and the couple itself. Together they constitute a holarchy, as the couple is larger than the two individuals "embedded" within it (no pun intended!). The integrity of each individual has to be endlessly negotiated with the integrity of the couple, seeking its own integrity or autonomy. Since ancient Greek times we have quipped that the problems of couplehood are that "you can't live with 'em and you can't live without 'em." There is a creative tension endlessly demanding resolution toward the dynamic balance of mutual consistency. As I see it, holons that do not establish mutual consistency with their surrounding holons will be isolated or destroyed unless they can kill or incorporate the holons in question.

In ancient India, a creation myth had it that the first wavelet formed in an endless sea of milk and ever after was torn between its love of self and its desire for merger. It seems to me that this very fundamental process of negotiating mutual consistency drives both human social relations and all evolution, which I see not as a neo-Darwinian process of random mutation and selection, punctuated or otherwise, but as an intelligent improvisational dance, about which we must say much more to make it credible.

While we are describing the processes of the living entities we now call holons in terms such as autonomy, holonomy, and mutual consistency, let me also mention the economy *(oikos + nomos)* and ecology *(oikos + logos)* of living systems, which is to say literally, in Greek, the rule and the organization of the "household." In our world at present we actually debate whether there need be a relationship between ecology and economics, and if so, what it should be. If we understand these original meanings and look at living entities as having fundamental patterns of process, we can easily see why we should not have separated their organization from their rule in our conceptions of human society or "household."

HARMAN:

I would like to go back to your thought: What if biology is taken as the root science? What if we start science by looking at organisms and other wholes, and recognize from the outset that the whole is not adequately understood in terms of its parts alone?

If there is anything we thought we could count on, it is the reliability of the "scientific method" for discovering "truth." However, there are too many things in the world that don't seem to fit with this view from reductionistic science—that is, the view that physics is the "queen of the sciences." We need an epistemology that is potentially capable of taking into account the amazing instinctive abilities of animals, the mysterious puzzles of evolution, the wondrous forms of flowers—and above all, the mystery of human consciousness and the human spirit.

If we accept Wilber's ontological stance, a good many seemingly opposing views in Western thought become reconciled. From the level of the human-holon, the scientist looks mainly inward in the holarchy (at physiological and smaller levels); the mystic looks mainly outward toward the universe. Science and religion are potentially two complementary but entirely congenial views; each needs the other. In Western philosophy there have been three main ontological positions: the materialist-realist, the dualist, and the idealist. The materialist looks at the entire holarchy as made only of matter, while the idealist sees the *entire* holarchy as made up of something more than matter. The dualist tries to reconcile fragments of these two views. Each position offers a partial glimpse of the holarchic whole.

I find that this new ontological stance takes some living with to fully appreciate how successfully it resolves many of the time-honored puzzles of Western philosophy—the mind/body problem, for example, and free will versus determinism. Since everything is part of the one holarchy, if consciousness is found anywhere (such as at the level of the scientist-holon), it is by that fact characteristic of the whole. We can rule it out neither at the level of the microorganism nor at the level of the Earth, or Gaia.

Nor should we be nonplussed by evidence that individual minds may appear to be connected together in ways not explainable through physical fields or mechanisms. There is much evidence, some recent and some ancient, suggesting that the healing of one person's body may be affected by the state of another, remote person's mind—in one frame of reference it is spoken of as "remote prayer healing." In our accustomed scientific worldview it may seem a preposterous hypothesis; in a holistic biology there is no *a priori* reason to rule it out. Since everything is connected anyway, the more appropriate question is not how can remote healing occur, but why don't minds interfere with other bodies even more than they seem to?

At the level of the human holon, we find that being alive implies intention, creativity, playful experimentation, desire to learn, seeking ever greater levels of complexity and diversity, continually searching for "what works." But then we have no justification for insisting that these features are not also characteristic of the greater holon. Thus we should be surprised to find, as is Margaret Wheatley (1996), that we live in a world in which life wants to happen; that the universe seems to be alive, creative, experimenting all the time to see what's possible; that it is the natural tendency of life to organize, and to seek ever greater levels of complexity and diversity; that all of life appears intent on finding "what works"; that life organizes creatively and spontaneously around a "self"; that we live in a world that is truly cocreative, in that every "self" at every level is both part of the creation and of the creating.

Following through the kind of thinking this holarchic concept sets in motion can become pretty mind-blowing. If we seek a new set of metaphysical assumptions on which to base a truly adequate science of biology, this is the best candidate I can find. It's a radical shift from the view that gives us molecular biology and quantum physics, yet it includes that view.

Of course the science based on this—should we call it holarchic?—worldview will still insist on *open inquiry* and *public (intersubjective) validation* of knowledge. However, it will recognize that these goals may, at any given time, be met only incompletely. Taking into account how both individual and collective perceptions are affected by unconsciously held beliefs and expectations, the limitations of intersubjective agreement are apparent.

This epistemology will be *"radically empirical"* (in the sense urged by William James) in that it will be *phenomenological* or experiential in a broad sense (that is, it will include subjective experience as primary data, rather than being essentially limited to physical-sense data) and it will address the totality of human experience (in other words, no reported phenomena will be written off because they "violate known scientific laws"). Thus, consciousness is not a "thing" to be studied by an observer who is somehow apart from it; consciousness involves the interaction of the observer and the observed, or, if you like, the *experience* of observing.

Modern scientists have had a certain tendency to misuse the "principle of parsimony," also called "Occam's Razor." Fundamentally, this principle states that one should cut hypotheses to the minimum. For example, do not postulate a vital force in matter if the phenomena can be explained without it. But the misuse is when the "razor" is used not to trim the fat out of hypotheses but to cut out observations of "anomalous" phenomena in order that the hypotheses may stand. For example, some scientists would deny evidence of so-called "psi" phenomena because they imply powers of the mind that are "known" not to exist.

This adequate epistemology will be, above all else, humble. It will recognize that science deals with *models and metaphors representing certain aspects of experienced reality,* and that any model or metaphor may be permissible if it is useful in helping to order knowledge, even though it may seem to conflict with another model that is also useful. (The classic example is the history of wave and particle models in physics.) This includes, specifically, the metaphor of consciousness.

That last statement may sound strange; let me explain. It is a peculiarity of modern science that it allows some kinds of metaphors and disallows others. It is perfectly acceptable to use metaphors that derive directly from our experience of the physical world, such as "fundamental particles" and acoustic waves, as well as metaphors representing what can be measured only in terms of its effects, such as gravitational, electromagnetic, or quantum fields. It has further become acceptable to use more holistic and nonquantifiable metaphors, such as organism, personality, ecological community, Gaia, and universe. It is, however, taboo to use non-sensory "metaphors of mind"—metaphors that tap into images and experiences familiar from our own inner awareness. I am not allowed to say (scientifically) that some aspects of my experience of reality are reminiscent of my experience of my own mind—to observe, for example, that some aspects of animal behavior appear as though they were tapping into some supra-individual nonphysical mind, or as though there were in instinctual behavior and in evolution something like my experience in my own mind of *purpose.*

The epistemology we seek will recognize *the partial nature of all scientific concepts of causality.* (For example, the "upward causation" of physiomotor action resulting from a brain state does not necessarily invalidate the "downward causation" implied in the subjective feeling of volition.) In other words, it will implicitly question the assumption that a *nomothetic* science—one characterized by inviolable "scientific laws"—can in the end adequately deal with causality. In some ultimate sense, there really is no causality—only a Whole evolving.

It will also recognize that prediction and control are not the only criteria by which to judge scientific knowledge. As the French writer Antoine Saint Exupéry put it, "Truth is not that which is demonstrable. Truth is that which is ineluctable." In other words, the unquestioned authority of the double-blind controlled experiment is thrown deeply into question.

This epistemology will involve recognition of the inescapable role of *the personal characteristics of the observer,* including the processes and contents of the unconscious mind. The corollary follows, that to be a competent investigator the researcher must be *willing to risk being profoundly changed* through the process of exploration. Because of this potential transformation of observers, an epistemology that is acceptable now to the scientific community may in time have to be replaced by another, more satisfactory epistemology by new criteria, for which it has laid the intellectual and experiential foundations.

SAHTOURIS:

Brian Arthur at the Santa Fe Institute is a rare scientist in recognizing that science does not work by deduction "but mainly by metaphor" (M. Mitchell Waldrop, 1992, p. 327), and that the choice of the appropriate metaphors is thus the main work of the Institute in its study of complexity (p. 332). Appreciative of this acknowledgement that choice of metaphor is central to science, I gave a colloquium at the Santa Fe Institute in response to it—on the subject of new metaphors, such as Gaia for the Earth. I really support your idea that we must take the limitations off this choice of metaphors while we are in such major transition.

Thomas Kuhn has even suggested that we suspend standard categories such as "real/unreal" and "scientific/non-scientific" in looking at new data to find new patterns, new ways of ordering, and new categories. Our holarchic model is a way of ordering things in the least prejudiced way we can find—simply observing that living entities occur within each other and are interrelated in their processes, in fact are processes in themselves. We can then look at their observable features from this holistic perspective, and as you suggest, refuse to reject anything we see in them on the grounds that it does not fit the conventional paradigm of what is real.

Our current neo-Darwinist account of evolution is intellectually and spiritually so unsatisfying that one can easily see why people flock to creationist alternatives, however dangerous they are socially in their fundamentalist politics. Humanity desperately needs an intelligent account of its own and other species' evolution.

That is exactly what you are calling for in saying we need "an epistemology that is potentially capable of taking into account the amazing instinctive abilities of animals, the mysterious puzzles of evolution, the wondrous forms of flowers—and above all, the mystery of human consciousness and the human spirit." Reducing biology to chemistry and physics has indeed held this up, because it has taken living phenomena down into the realm of the nonliving and tried to explain them in its terms. Yet, in all fairness to physics, it had come much farther than the models to which biologists were reducing their phenomena—recall the two-train metaphor I mentioned earlier.

Someone pointed out that by using Western science to study consciousness, we arrived at the brain; then studying the brain we arrived at its physiology; studying its physiology we arrived at its chemistry; and studying the chemistry we arrived at physics, only to discover that in the study of physics we arrived at consciousness! What interests me at present is the confluence of one strand of thought in both biology and theoretical physics: that consciousness pervades everything and that therefore the entire universe must be alive.

HARMAN:

Related to that, I'd like to explore further the issue of intentionality in living systems and in the evolutionary process. I recall biologist Edmund Sinnott's argument, decades ago, that the insistent tendency among living things for bodily development to reach and maintain as a "goal," an organized living system of a definite kind, and the equally persistent directiveness or goal-seeking that is the essential feature of human behavior and mental activity, are fundamentally the same thing.

This was in his 1955 book *The Biology of the Spirit*. I don't believe that the question he raised then has yet been satisfactorily answered: What is the relationship between the self-organizing, apparently goal-seeking tendencies found in all living organisms and those aspirations and yearnings which mold human behavior and comprise the human spirit? Perhaps the question could not have been dealt with a few decades ago. The spectacular achievements of modern biology have been associated with the assumption that all the processes of life can be interpreted finally as simply physical and chemical ones. Nothing else appeared to be needed for the impressive developments in biotechnology. A strong faith had developed that the remaining mysteries would yield to the same sorts of explanation. The prevailing attitudes did not encourage taking seriously any question phrased in terms of spirit.

However, by the 1990s numerous challenges had arisen to the adequacy of a positivistically framed biology. Revised interpretation of early evolution had suggested that even the most primitive forms of life, the prokaryotes (from the Greek for "before a kernel," or a nucleus), responded to drastic changes in their environment in ways that imply something like creative problem solving. New research on the great radiations of macroevolution indicated that new phyla, classes, and orders have appeared relatively suddenly, especially following the major extinctions, in what seem much like outbursts of creativity. Most important, a long-standing and persistent sense that something important was missing from the picture of the world emerging from reductionistic science had led to growing interest in the possibility of a more adequate scientific account of consciousness. The time seems right for once again raising Sinnott's question.

So what about the human spirit? If through the biological sciences some light is shed on the evolution of spirit, does that reduce the human spirit to something less? Or does it elevate the way we view biology?

Basically, Sinnott's hypothesis equated the "biological goals" of growth and development in plants and animals with the goal-seeking characteristic of human behavior and mental activity. He anticipated objections to his hypothesis; as he said, "A conscious purpose in mind and an embryonic process in a developing tadpole appear at first to be so far apart that to regard them as fundamentally the same sort of thing may seem indeed preposterous. . . . Scientists have fought so hard to keep the insidious idea of purpose *out* of biology that they will not readily assent to a concept that puts this fighting word back at the very heart of the life sciences. . . . Of one thing we can be sure: there is inherent in the living system a self-regulating quality that keeps it directed toward a definite norm or course, and the growth and activity of the organism takes place in conformity to it."

One of the examples used by Sinnott was that of starting a new plant from a "slip," or cutting of an existing living plant. A small cutting of the parent plant, removed from its stock and placed in the ground, restores the parts it lacks and becomes a whole new plant. If, for example, one cuts off a short segment of willow stem in the spring and wraps it in damp moss or otherwise provides a favorable environment, roots will begin to grow from it, and buds will swell and grow. But the roots will come from what was once the lower end of the twig, the end that was cut off from the parent tree. And the buds or shoots that push out come from the other end. Whether a given cell will start to create buds or roots depends on whether in the selection of the segment the cell ended up at the upper end, or at the lower end. It makes no difference if the twig is kept horizontal or upside down; roots will still grow from what was the end attached to the parent plant, and shoots will grow out at the other. Even a tiny portion of the original organism contains within it the knowledge of how to create a complete plant. Somehow "there must be present in the plant's living stuff, immanent in all its parts, something that represents the natural configuration of the whole . . . a 'goal' toward which development is invariably directed."

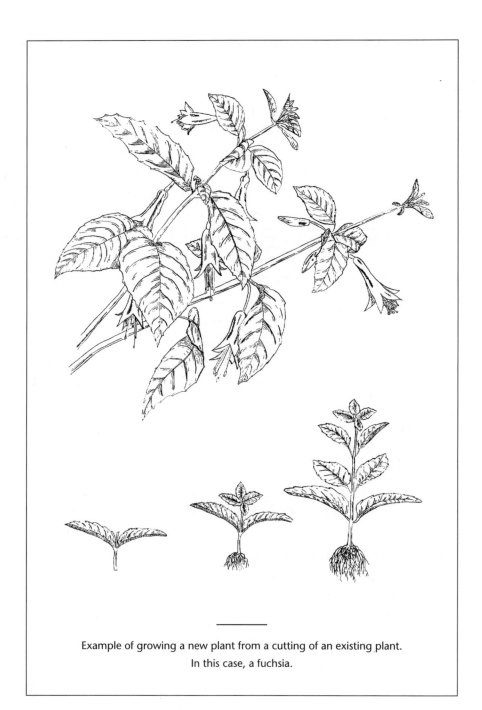

Example of growing a new plant from a cutting of an existing plant.
In this case, a fuchsia.

Such examples barely sample the wide range of biological puzzles that have resisted satisfactory explanation through physico-chemical processes alone. Because of the specialization that has almost universally characterized academic research, these puzzles have been considered separately, by different scientists, from another scientific puzzle; namely, what to do about consciousness. Consciousness research, it turns out, seems to entail a different epistemology, a different way of testing what is to be designated "scientific knowledge," if consciousness is to be satisfactorily included in the scientific accounting of human experience. In a way, what we are doing with the "holons within holons" approach is to bring the two together. It may seem unlikely that by combining two major perplexities we would gain insight rather than compound puzzlement, but let us see.

Sinnott raised his issue in the 1950s, and I believe there has not yet been a satisfactory response. As we have noted, the interpretation of biological processes in terms of underlying physical and chemical phenomena has long dominated the scene. But there appears now to be a new openness with regard to exploring such questions as that raised by Sinnott.

The semblance of purpose in animal behavior and plant tropisms, in instinct and in evolution, is commonly explained away as simply evidence of our innate tendency to anthropomorphize. But perhaps we are entering a new era in which such questions will be looked at differently.

SAHTOURIS:

As you say, most biologists have found it easier to ignore Sinnott's questions or to deny the observed teleology, than to look at the whole paradigm of biology anew. I assume we have adequately dealt with mechanomorphism as secondhand anthropomorphism. I like Walter Pankow's observation that it takes a living system to know a living system. If we ourselves were not conscious, we could not even imagine anything else to be conscious and could not discuss any of these issues with each other.

The hardest question you asked is whether we can shed some light on the evolution of the human spirit through the biological sciences, and whether that would reduce the human spirit to something less, or elevate the way we view biology. This question is still difficult for me because I cannot resolve our perception of spirit into a single explanation. As a scientist, I seek to do so, but it may be better for us to adopt Paula Underwood's (1991) intelligent native approach to such matters: to hold in mind simultaneously alternative views of something we wish to explain but cannot yet, and award them different weight as evidence comes in.

I'm not sure what is meant by the term "spirit"; it will be enough for me if the biological sciences come to see consciousness as a fundamental characteristic of nature, as my own concept of spirit is simply the most refined aspect of consciousness. Not so long ago, I still held the view that consciousness and intelligence are emergent phenomena in evolution, but spending time in native cultures has given me a broader perspective. Now I see consciousness and intelligence permeating the universe from as far back as we can see any patterns forming. That also makes it possible that consciousness, or spirit if you prefer, preceded the known material universe, as in the well-known interpretation of quantum theory. Eugene Wigner, for example, has stated outright that consciousness must be introduced into the laws of physics.

This is the view that I've come around to, radical as it will still seem to many biologists. Research in material biology is giving us incredibly detailed views of what goes on, down to the molecular level, but none of it can explain *why* these marvelously

complex events take place. Once we assume that intelligent consciousness transforms itself into matter, the "why" of it becomes
more accessible. This is not a dualistic view any more than is our
understanding of energy-matter transformations in physics. It
simply sees consciousness as the first stage, transforming itself in
turn into electromagnetic energy and then into matter.

In some traditions I believe spirit is equated with the nonmaterial field variously called the auric body, the soul body, or the
subtle body. I would like to refer later (Chapter Five) to recent
research on some measurements of physical quantities that
appear to be associated with the "subtle bodies," long known in
other cultures. If such research leads us to incorporate spirit into
biology, I would hope that the human spirit will not be dimin-
ished, but that biology will be elevated.

Your examples of teleology clearly come from different layers,
or "onionskins," of planetary holarchy, as we have been discussing it: the problem solving of microbes, the plant cutting
that forms a whole new plant, the human being experiencing his
or her own mind in goal-oriented behavior, and the outbursts of
creativity in planetary evolution after mass extinctions.

What that says is that we can find intentional, goal-oriented, or
teleological behavior wherever we look in nature. The truly strange
thing is how effectively we can be taught in graduate school to
deny our own observations, to call what we see something else, to
pretend at first and then to convince ourselves that there is nothing purposive going on. I was even taught that the mind does not
exist, not even in humans—that it is an unnecessary hypothetical
construct. Everything, we were told, could be explained in the
reductionistic terms of molecular chemistry (at its level) or in terms
of behavioral reinforcement (at another level). Wholes were no
more than the sum of their parts, and the miracle of science was

held to lie in our ability to see the tiniest parts. That physics went that route before biology and stumbled back upon consciousness was not discussed or was dismissed as something unique to quirky physics and its mathematical games. Then, at last, we were awarded degrees and felt proud that we had become real scientists who saw more truly than those who are not. And indeed, we believed it! Recently I ran into a note I had scribbled some years after finishing graduate school: "I sometimes think I was awarded a Ph.D. for successfully denying my humanity—behaving mechanically, dissecting nature without feeling or emotion."

Purpose, spirit, even mind were not part of the Western scientific worldview or, within it, of biology and experimental psychology. Only science itself had a purpose—to reduce everything down to something so simple that we could control it, that is, to use our information about the world to help us transform it to human use. Scientists are still, of course, quick to protest that science is an absolutely neutral fact-finding endeavor, but only because of its usefulness is science so well funded by public monies, especially from military budgets and by private corporations and their endowments.

If science did not deny consciousness and intelligence in nature, things would look very different. We could no longer see nature as an array of resources for human use; we could no longer see ourselves as the only intelligent beings. We humans would not be in full control and our technological solutions to problems would often come into question.

It is of great interest to watch how non-scientists are reacting to the technological control over nature coming out of science.

For example, people are turning in droves to traditional forms of growing food and healing by hands-on body work—"shamanic healings" and herbal remedies that often come from indigenous peoples' knowledge. All these forms of healing respect the intelligence of the body to heal itself with minor assistance from natural sources rather than dramatic technological interventions. Even if they do not always work, they seem to do well enough, and scientists are already hard pressed to deny their efficacy.

In 1974 I had the opportunity to go to China and see the integration of traditional and Western medicine. The scientists and medical personnel were working very hard to determine which system worked best for which cases or when both could be effectively combined. In fact, their acupuncture research at that time was already far more scientific from a Western perspective than is ours in the West today. When acupuncture did make Western inroads, it was through the alternative healing professions, not through science laboratories. That meant it was rushed directly into practice, without even looking at what Chinese scientists had discovered about it neurologically from their laboratory research with animals.

The point is, the Chinese were already working on a very interesting integration of traditional knowledge and formal research science, something like what Mae-Wan Ho proposes as a modern/indigenous science (1988), and which my own experience of native sciences strongly supports. But let's leave that discussion for later on, so we can continue from your discussion of Sinnott's work and get to some of the other new findings in biology and the questions raised by them.

Interlude One

The Amazing Prokaryotes

This history of the amazing prokaryotes is such an exciting story that it deserves telling in more detail. The first life-forms appear to have been microbes with permeable lipid walls; as these walls evolved to more rigid forms we have the first bacteria—prokaryotes; that is, cells without a nucleus. This was 3.5 billion years ago, and for the next two billion years Earth was inhabited solely by prokaryotes. (The fascinating story of the ingenious prokaryotes and their later alliances to form more complex biological organisms is well told by Lynn Margulis, who made many of the original discoveries.)

The important events in this early evolution have to be hypothesized, since they left no fossil record. First life probably resembled the minimal organisms that exist today. Minimal life seems to contain enough DNA to direct metabolism to a limited extent. However, there are not enough genes to take care of making the amino acids, nucleotides, vitamins, and enzymes needed for life processes. We have to assume that first life absorbed its components directly from the environment. (Contemporary forms of these minimal kinds of bacteria are parasites that get what they need from the organisms in which they live, but the earliest life forms had no such friendly environments. These elementary life forms used the biochemicals that had accumulated due to exposure of chemical mixtures to ultraviolet light and lightning in the absence of oxygen.)

Of course, this chemical feast couldn't last forever. Such free nutrients were quickly depleted as the microbes ceaselessly ate, grew, and divided to produce more microbes. Ordinary hazards of the environment—temperature variations, the quality and quantity of sunlight, the concentration of salts in the water—all acted to diversify the populations of microbes in different places. In the face of starvation, there began to develop a variety of successful new bacteria with new metabolic strategies by which food and energy could be extracted from various raw materials.

One of the first innovations enabled cells to use sugars and to convert sugar to ATP (adenosine triphosphate) energy. This key chemical is used by all living cells without exception as a carrier for energy; it has been called the general "biological currency" of stored chemical energy. It is the energy source for many basic metabolic reactions as well as for more specialized energy-requiring reactions, such as muscle contraction. It was presumably present in the primitive, oxygen-free atmosphere; so also were sugars from which it could be made. Thus these cells lived off chemicals they could readily find in the earth. They developed various sugar breakdown processes, known as fermentation. The lower-energy byproducts of sugar—alcohols and acids—were then excreted as waste.

Some of these "fermenters" began with sugars (glucose, sucrose) or with carbohydrates (cellulose, starch). Others began with simple nitrogen-containing compounds, such as amino acids. Some ended up with carbon dioxide and ethanol, like the bacteria that ferment the ingredients for wine and beer; others with lactic acid, like those that sour milk and ripen some cheeses; still others with acetic acid and ethanol, like those that form in sewage or make vinegar. The fermentation process generally gives the cell a few molecules of ATP for every food molecule broken down.

Since the waste products of fermenting bacteria—certain acids and alcohols—still contain energy, fermentation isn't entirely efficient. In time, other microbes evolved that ate the wastes of fermenting bacteria. These new bacteria broke down the wastes, deriving more carbon and energy from them. These processes still go on where the quantities of light and oxygen are low—in swamps and lake muds, in tidal flats, in animal guts, and in standing puddles—one fermenter's food being another's waste.

As the prebiotically produced organic compounds were gobbled up and became scarce, other ways of creating ATP developed. One group of bacteria, known as desulfovibrios, learned to "eat" sulfate, emitting noxious sulfur gases. They generate ATP during the conversion of sulfate to sulfide. They emit hydrogen sulfide, the odoriferous gas that gives salt marsh mud and some hot springs their "rotten egg" odor.

That was followed by another, extremely important "invention": photosynthesis. When any molecule absorbs light, its electrons are boosted to a higher energy state. Under certain conditions, in bacteria that contain a kind of molecule called a porphyrin ring, light energy can be retained and converted to ATP. ATP energy is then used for movement and synthesis, such as conversion of carbon dioxide from the atmosphere into the food and replicating carbon compounds needed to self-maintain and grow. Early photosynthesis was different from that found in plants today. These sun-loving bacteria used light energy and hydrogen sulfide to create ATP, excreting sulfur; the green-sulfur and purple-sulfur bacteria still carry on in this way.

Grass-green and particularly blue-green cyanobacteria brought in another innovation; they learned to "eat" carbon dioxide, giving off oxygen as a waste product. This is the form of photosynthesis eventually responsible for the success of the plant kingdom. As this oxygen "polluted" the environment, a new type of oxygen-breathing bacterium arose. Thus was born another metabolic path by which ATP was created—aerobic respiration, the most efficient of all.

A new kind of blue-green cyanobacteria developed the ability to split the water molecule, extracting hydrogen to combine with carbon dioxide from the air to make sugars and other organic food chemicals, and excreting oxygen as waste. The blue-green bacteria were so prodigiously successful that they literally covered the earth, wherever there could be found together water and sunlight. Due to their ubiquitousness and their appetite, the earth became awash with oxygen; the oxygen content of the atmosphere rose from around 0.0001 percent to 21 percent! But uncombined oxygen was toxic to most of the life forms then existing. Thus about two thousand million years ago there was a worldwide oxygen-pollution crisis.

However, the clever cyanobacteria were up to the challenge. They "invented" a metabolic system that fed off of the very substance that had been a deadly poison—aerobic respiration, the breathing of oxygen, wherein the destructive energy of oxygen is used to break up food molecules and thereby free both their parts and their energy for use. This is essentially controlled combustion. Organic molecules, are broken down to yield carbon dioxide, water, and a great deal of energy. Whereas fermentation typically produces two molecules of ATP for every sugar molecule broken down, the respiration of the same sugar molecule utilizing oxygen can produce as many as thirty-six. This was such a spectacularly successful leap in technology that the cyanobacteria went on an evolutionary binge. They exploded into hundreds of different forms, the largest being about eight-tenths of a millimeter in diameter. They spread throughout the environment, from cold marine waters to hot freshwater springs. They literally covered the Earth.

The more one thinks about this story of the prokaryotes' colonization of the entire Earth's surface, the more amazing it seems. During the almost two billion years when they were the Earth's sole inhabitants, the prokaryotes continuously transformed the Earth's surface and atmosphere, organisms and environment evolving together. They "invented" all of life's essential, miniaturized chemical systems: fermentation, photosynthesis, oxygen breathing, and the removal of nitrogen gas from the air. They also experienced crises of starvation, pollution, and extinction—all before the appearance of the first nucleated cells (eukaryotes).

Lynn Margulis was one of the key investigators who put this story together (1986). According to her, throughout this early phase of Earth's history there appear to have been two major mechanisms of evolution—mutation of DNA and bacterial genetic transfer. (A third, symbiotic alliance, doesn't come in until later, in connection with the evolution of eukaryotic cells (cells containing a nucleus).)

Mutation of DNA is not only random; in some circumstances it appears to be purposeful. In the case of the prokaryotes, as they ran low on one kind of food they seem to have "invented" new ways of metabolizing other foodstuffs. When the evolutionary role of "purposeful mutation" was first proposed, it seemed a preposterous notion, another instance of the human tendency to anthropomorphize nature. It does seem an absurd thought, given our minified conception of a microbe's capabilities, that it could appear so clever.

However, subsequent research seems to substantiate the idea of purposeful mutation. If bacteria are placed in a medium with nutrient molecules too large to pass through the pores of its membrane, they may mutate to make the membrane more permeable. Bacteria grown in a salty medium become better able to survive and reproduce in seawater. John Cairns and colleagues (1988) found that a strain of the bacteria *Escherichia coli* unable to digest lactose became able, through "purposeful mutation," to metabolize it when they had to. (Not only did the bacteria have a predisposition to mutate for a lactose-digesting enzyme; more mutations were induced by placing the bacteria in a lactose solution. No mutations to revert to the capacity to use lactose seemed to occur as long as other nutrients are available, but they appeared when the medium contains only lactose.) In another experiment, Barry Hall (1988) found that when E. coli in a solution of salicin was given little other nourishment, it underwent two otherwise rare mutations together at a rate thousands of times higher than in normal growth, making itself able to use the salicin. He suggested "that cells have some means of recognizing what would be an advantageous mutation and increasing the chance that it occurs."

Bacterial genetic transfer is the other early mechanism of evolution. Prokaryotes routinely transfer bits of genetic material to other individuals. Each bacterium at any given time has the use of accessory genes, visiting from sometimes very different strains, which perform functions that its own DNA may not cover. Some of the genetic bits are recombined with the cell's native genes; others are passed on again. As a result of this ability, all the world's bacteria essentially have access to a single gene pool and hence to the adaptive mechanisms of the entire bacterial kingdom. Margulis describes this in a startling way: "By constantly and rapidly adapting to environmental conditions, the organisms of the microcosm support the entire biota, their global exchange network ultimately affecting every living plant and animal. Human beings are just learning these techniques in the science of genetic engineering, whereby biochemicals are produced by introducing foreign genes into reproducing cells. But prokaryotes have been using these 'new' techniques for billions of years. The result is a planet made fertile and inhabitable for larger forms of life by a communicating and cooperating worldwide superorganism of bacteria." (1986, p. 18–19)

As a consequence of these two mechanisms, there evolved various forms of prokaryotes, adapted to different environments and ways of life. These included sulfur bacteria (capable of feeding off the sulfur compounds found in volcanic vents, etc.), methanogens, halophiles (living in salt brine), cyanobacteria (blue-green algae), myxobacteria, gram-positive bacteria (e.g. bacillus), gram-negative bacteria (e.g. *Escherichia coli*), purple photosynthetic bacteria, and two kinds of bacteria that became mitochondria and chloroplasts living within nucleated cells. Those last two are especially interesting. Let's come back to them.

Quoting Margulis again, by "1.5 billion years ago most of biochemical evolution had been accomplished. The Earth's modern surface and atmosphere were largely established. Microbial life permeated the air, soil, and water, cycling gases and other elements through the earth's fluids as they do today. With the exception of a few exotic compounds such as the essential oils and hallucinogens of flowering plants and the exquisitely effective snake venoms, prokaryotic microbes can assemble and disassemble all the molecules of modern life." Prokaryotic biota had stablized atmospheric oxgen more or less at its present level of 21 percent. "Growing, mutating, and trading genes, some bacteria producing oxygen and others removing it, they maintained the oxygen balance of an entire planet." This "cybernetic control of the Earth's surface by unintelligent organisms calls into question the uniqueness of human consciousness." (p. 113)

One of the most important developments was yet to come; the evolution of the first eukaryote, or nucleated cell, possibly as long as 2.2 billion years ago. The eukaryotes comprise all of the remainder of the single-celled and multicelled living world, except for the prokaryotes. All eukaryotic cells share a typical compartmentalized internal design, their DNA and the machinery for its transcription into RNA being sequestered in a distinct membrane-bound nucleus. Aerobic respiration and, in green plants, photosynthesis, are carried out in specialized organelles—mitochondria and chloroplasts, respectively. (Organelles are to cells as organs are to multi-celled organisms.)

Animal cells are distinguished from plant cells by two features. The first is their lack of chloroplasts, which, containing light-activated chlorophyll, are the organelles of photosynthesis. The second difference is the absence of a relatively rigid cell wall of cellulose, a plastic-like carbohydrate. (Fungi are not plants. They have a cell wall typically composed of chitin, a fibrous carbohydrate, and do not include chloroplasts.)

Mitochondria, found in nearly all animal, plant, and fungal cells, are responsible for respiration; they produce the energy-transferring molecule ATP (adenosine triphosphate). Mitochondria are tiny membrane-wrapped inclusions that lie outside the nucleus and have their own genes composed of DNA. Both chloroplasts and mitochondria have lamellae or platelike invaginations of an inner membrane.

A typical cell also contains tens of thousands of ribosomes, tiny assemblages of protein and RNA (ribonucleic acid) that string amino acids into peptides and proteins. The size and shape of ribosomes in eukaryotes and prokaryotes are different.

Through genetic transfer, similar to that which occurs in prokaryotic cells, visiting genetic bits can move into the genetic apparatus of eukaryotic cells. But an even more remarkable mechanism of cell modification appears to be responsible for the ability of eukaryotes to create multicelled plants and animals—namely, symbiotic alliance. Both mitochondria and chloroplasts are now generally understood to have come about through the symbiotic alliance of particular

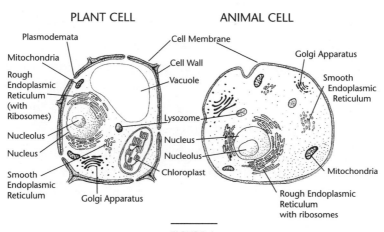

PLANT CELL   ANIMAL CELL

Plasmodemata
Mitochondria
Rough Endoplasmic Reticulum (with Ribosomes)
Nucleolus
Nucleus
Smooth Endoplasmic Reticulum
Golgi Apparatus

Cell Membrane
Cell Wall
Vacuole
Lysozome
Nucleus
Nucleolus
Chloroplast

Golgi Apparatus
Smooth Endoplasmic Reticulum
Mitochondria
Rough Endoplasmic Reticulum with ribosomes

FIGURE 4:

Comparison of a generalized plant and animal cell. The two major differences are the presence of a rigid cell wall and chloroplasts (for photosynthesis) in the plant cell which are lacking in the animal cell.

bacteria that had learned useful skills—energy transformation and photosynthesis, respectively—with nucleated cells, thus imparting to them these important abilities.

And so we come to the third mechanism of early evolution, *symbiotic alliance.* Descendants of the bacteria that breathed oxygen in primeval seas three billion years ago now reside in cells of animals, plants, fungi, and protists (unicellular life forms) as mitochondria. Sometime in the distant past, these ancient bacteria appear to have taken up residence inside other cells, providing waste disposal and oxygen-derived energy in return for food and shelter. Unlike the cells in which they reside, mitochondria reproduce by simple division. Like most bacteria and unlike the complicated reproduction of the rest of the nucleated cell, mitochondria pinch and divide in two to reproduce, usually doing so at different times from each other and from the rest of the cell.

The chloroplasts that are found in all plant cells, where the process of photosynthesis takes place, are another example of symbiotic alliance, the merging of organisms into new collectives. Chloroplasts appear to be descendants of the early bacteria that learned to derive chemical energy from the sun. Some time in the distant past they evidently "made a deal" with some eukaryotes to take up residence and provide that useful ability in return for shelter.

Yet another example of symbiotic alliance is the merging of spirochetes with eukaryotes to create the cilia used for locomotion of single-celled organisms and to produce a transport of fluid in multicelled organisms. It strains the imagination to try to picture these remarkable examples of cooperation between microorganisms as having just happened by chance.

Now just to summarize all this: In the course of evolution, three ways developed of making the energy-storing molecule ATP. First was the fermentation bacteria, which made it from naturally occurring large molecules. Second was through "inventing" photosynthesis, and learning to trap solar energy and turn it into ATP energy through light-sensitive chemicals, including, especially, chlorophyll. These "blue-greens" flourished, but in the process they created great amounts of a poison, namely oxygen. The third and most efficient way of creating ATP was "invented" by bacteria that developed the process of aerobic respiration, wherein the destructive energy of oxygen is used to break up food molecules and thereby free both their parts and their energy for use. The next great evolutionary leap was the eukaryotes, complex single-celled organisms that evolved from "bacteriological cooperatives" and are the basic building blocks for higher organisms.

However one chooses to talk about the hierarchical ordering of atoms → molecules → organelles → cells → tissues → organs → organisms → societies → Gaia, the autopoietic tendency appears to be present at least from the eukaryotic cell up. In perhaps a modified sense it seems to show up in bacteria and organelles—in the amazing prokaryotes. In viewing "cells as microbial communities" that emerged over evolutionary time from ancient bacterial alliances; in the microbial mat communities, termite colonies, lichens, etc.; in the single global system, Gaia—in all these examples of "ecological individuals," exhibiting as they do some sort of basic cooperative tendency—the autopoietic impulse can be observed and described. A profound and pervasive underlying mystery remains.

CHAPTER
TWO

Some Biological Enigmas

The metaphor of natural selection derives from the dominant socio-economic ideology of the Victorian era, now rejected by nearly all of humanity. The mechanistic conception of life which it inspires is equally outmoded and inappropriate. Why should we cling to this metaphor when it can serve no other purpose than to reinforce those prejudices which gave it birth?

—Mae-Wan Ho (1988)

SAHTOURIS:

Let's go back to the question of teleology in biology, as it is such a central enigma. Take, for example, the remarkable ability of the prokaryotes, not only to make use of whatever they could find in the environment to derive the energy needed for living, but also to reinvent themselves with their accumulation and easy trading of genes when things got tough for one reason or another. Most impressive of all, as time went on, their shift from a competitive, exploitative lifestyle to one of cooperation that made all further evolution, from eukaryotes to ourselves, possible.

This cooperation we see emerging in the process of creating the eukaryotic cells, the nucleated single-celled creatures, implies something resembling consciousness or intentionality. It is not enough simply to take the Darwinist position that the dying out of failures adequately explains the successes. That would be like saying a human inventor, in throwing out his failed designs, has simply stumbled on his inventions.

When you are trying to explain very ancient evolutionary events for which there is almost endless time in which to account for changes, it seems easier to play the game of Darwinism and assume that things *could* proceed by accident, a mutation at a time. But when scientists, such as Lynn Margulis and Mae-Wan Ho, began watching live modern bacteria, not to mention larger organisms, and saw them responding to changes in the environment by appropriate mutations in very short order, it became impossible to deny intelligence and intentionality.

Yet despite the literature and live conferences by way of which biologists share the information they glean through research, it seems to take a long time for them to believe each other's good evidence, especially if political climates interfere.

Thus, for example, it has been known since the first few decades of the twentieth century that the famous Weismann barrier between reproductive and body cells (meaning that the organism's experience cannot affect its genes) did not hold because the germ cells of most animal phyla and of all plants develop directly from somatic cells, meaning that somatic mutations in response to the environment could be passed on to subsequent generations. This supported the Lamarckian inheritance theory popular among Soviet scientists, so of course it had to be denied in the Darwinian West.

Our computer models of nature can depict complexity to a certain degree, but they are notoriously unsatisfactory in demonstrating evolution. As Koestler suggested, and I emphatically agree, genetic mutations as copy errors or damages are more likely to be intelligently repaired than they are to be the source of evolutionary change. Evolution simply cannot proceed unless individuals within species can intelligently draw upon their accumulated reserve gene pool (and perhaps invent new genes on the spot) when they face critical situations or want to play on into new complexity.

More and more biologists are concluding that genes—DNA sequences that are blueprints for individual proteins—cannot account for the morphogenesis of organisms. Barbara McClintock (1984) found evidence that evolutionary change can be brought about by a natural increase in DNA transposition under stress. Her work on transposable elements was corroborated and elaborated, so it is now clear that DNA reorganizes itself and trades genes with other cells via viruslike elements known variously as transposons, and as retrotransposons and retroviruses with their RNA to DNA transcriptions—violating the old view that messages only come from the nucleus as DNA to RNA transcriptions. Some of such transfers appear to be "evolutionarily related to 'free-living' viruses," according to H. M. Temin and W. Engels (1984). Retroviruses are known to infect across species and enter the host's germline. Genes can thus be amplified, shuffled, and traded within and among somatic cells as well as with other individuals and species.

It would appear that organisms have a rich and varied set of tools for restructuring themselves at genetic levels. This relatively new view of organisms as proceeding intelligently in their own evolution is evidenced in the use of vocabulary such as "editor genes" when the entire explanation is sought within the DNA, or "morphogenetic fields," introduced by C.H. Waddington (1961) and elaborated by Rupert Sheldrake (1981), when deeper sources of such change are sought beyond DNA.

It sometimes seems that nature is being very obvious with its intelligent intentionality, perhaps to alert us blinded human scientists before we go farther astray in our blundering efforts to control her. To wit, the following report from the "I Told You So Department" of the *Los Angeles Times,* March 7, 1996 : "In an unsettling turnabout of modern farming's biological warfare against weeds, a team of Danish researchers reported today that some genetically engineered plants designed to withstand herbicides can pass those new genes to nearby weeds, which in turn become resistant to chemicals meant to eradicate them. Experts said the finding is the first confirmation of what many critics of the new biotechnology have long suspected—that new traits introduced into genetically engineered crops [rape in this case] in some instances can be inherited by nearby weeds and other wild plants that belong to the same general family."

Such reports will not be surprising to a group of English biologists including Brian Goodwin, Mae-Wan Ho, Jeffrey Pollard, and mathematician Peter Saunders, who have clearly formed a very interesting group at the forefront of the opposition to Darwinism and neo-Darwinism, based on strong laboratory evidence, with considerable support among biologists in the United States. Among the events at which various members of this group gathered, and which I was privileged to attend, were three Camelford Conferences on the implications of the Gaia thesis, convened by ecologist Edward Goldsmith in Cornwall in the late 1980s, with James Lovelock and Lynn Margulis present for the most part. All of their work is highly relevant to our discussion and may do much to bring biology into greater prominence in the near future.

Mae-Wan Ho and Jeffrey Pollard (1988), for example, have reported the rapid restructuring of genomes in response to stress, observing this process in many different species from microbes to plants and animals. This can bring about, as Pollard says, "dramatic alterations of developmental plans independent of natural selection," which itself may "play a minor role in evolutionary change, perhaps honing up the fit between the organism and its environment."

We are now in our fifth generation of antibiotics—which literally means "antilife agents" in Greek—because the attacked bacteria are fighting back by evolving into new resistant strains. In the *Los Angeles Times* of February 20, 1996, it was reported that the antibiotic vancomycin, considered "the big gun against troublesome bacteria that ran roughshod over less potent antibiotics," and "the ultimate protection against a wide range of bugs that could cause fatal infections," was no longer effective. No one expected, said the *Times,* that "a bacterium would outsmart every antimicrobial, including mighty vancomycin, the so-called drug of last resort. First reported in Europe in 1988, van-comycin-resistant enterococcus, a gastrointestinal bacterium, has spread methodically westward; sporadic cases began crop-ping up in California in 1994. And while vancomycin-resistant enterococcus is no Andromeda strain, its presence signals the ominous dawn of an era in which infectious diseases that once were put down easily with antibiotics could become incurable. Outbreaks in Africa of the deadly Ebola virus helped awaken the public to the threat of incurable infectious disease. But experts say that what the American public, and even their physicians, may fail to realize is that the problem is now at hand as com-mon bacteria find ways to survive an arsenal of overworked and misused antibiotics."

It is clear that the Darwinian timeframes of accidental muta-tion are not at work in these current examples. What new light does this shed on ancient evolution—for example, when the tasks of eukaryotic (nucleated cell) life were apportioned among the various prokaryotes (bacteria) that came to play the roles of chloroplasts and mitochondria? The intentional cooperation seen here is not naive anthropomorphizing; we are beginning to see that there is too much to be explained away unless something like awareness and intentionality are admitted. Later, when we get around to discussing indigenous sciences, we can see how other cultures have always known every part of nature to be intelligent and have developed beneficial, instructive communi-cations with them.

Clearly, besides the inconvenience and expense involved in such cases as the genetic engineering failures, our tampering with natural processes has very serious consequences for us all, possibly to the extent that the overall harm we do ourselves ends up exceeding the benefits. Farmers now have to use many times the pesticides per acre they used in the 1940s and are still finding increased rates of pest damage, while we consumers eat increasingly poisoned crops and the birds disappear, as Rachel Carson predicted. Visits to hospitals are increasingly dangerous in terms of infections picked up there.

Only a few years ago we watched as malarial microbes mutated in response to medication designed to kill them. They may have done it by the known process of gene amplification to increase the concentration of specific enzymes that make the microbe resistant to specific toxins. They then passed this "information" to other microbes as the mutants traveled from Asia to South America on human hosts, soon making the prevention and treatment of malaria vastly more difficult than it had been, around the world. All this happened far too quickly to be accounted for by Darwinian means.

HARMAN:

That matter of apparently smart, purposeful bacteria is certainly a key unsolved puzzle. It seems to me that so also is the development of form—morphogenesis. Most theories of embryonic development in the last seventy years have attempted to make something of the notion of positional information—the idea that the present location of a cell and its present activity provide most of the information on what it is to do next. Yet, as you said, the term "editor genes," for example, has crept into the language, implying the suspicion that there may be something more than mechanism. Morphogenesis remains one of the key enigmas; once consciousness enters the picture, the metaphor of a "morphic image" is suggested.

This concept of a guiding image is a fascinating thought. I think it might help us understand the major puzzles in evolution that seem to represent failures of the neo-Darwinist thesis, which some biologists are beginning to come to grips with in their own way. I want to talk a bit later about Levins and Lewontin's illuminating description of the dialectical nature of evolution of organisms and environment evolving together. And also about Brian Goodwin, who has an interesting approach with regard to the evolution and ontogeny of form. He claims that the information in the DNA is a necessary but not sufficient determiner of form. All of this casts considerable doubt on the sufficiency of the information in the genes (DNA) to fully define the organism.

SAHTOURIS:

Certainly everyone present at the Cornwall Gaia conferences would have agreed that DNA is a necessary but not sufficient determiner of form. Ho (1988) gives numerous examples of morphological change induced by environmental conditions during development, such as fruit fly larva exposed to ether at particular ontogenic stages and responding with predictable changes of form, or E. coli that cannot metabolize lactose quickly mutating two separate genes to produce a lactose-metabolizing enzyme when placed in environments where they must metabolize lactose to survive. Further, the exact same two mutations appeared in thirty-one of thirty-four replications of the experiment in different laboratories. She concludes, "It is the physiological state of the cell in one case, and the epigenetic system of the organism in the other, that organically 'selects' the appropriate response."

The use of that term "editor genes" by a few mainstream microbiologists actually understates the case that now emerges from research on the fluidity and flexibility of the genome and its changes throughout the lives of organisms. This fluidity and change in the phenotype's response to changes in its environment is as far as can be from the petty random processes of neo-Darwinism. It is not only indicative of a dialectic between individual organism and environment, but of the transcendent self-reference of entire ecosystems and of the planet as a whole. I think many of these enigmas will be cast in a very different light as we include consciousness and intelligence in the new paradigm.

HARMAN:

There are plenty of other puzzles to underscore the point. Consider the puzzle of recognition. A lot is known about how organisms recognize others of their same species, and how they recognize the opposite sex. The body's immune system can recognize tissue as being from itself or some other organism. White blood cells do an amazing job of recognizing invaders of various types. It's a little difficult to see how that ability can be explained through a "program" in the DNA.

But there are many other forms of innate behavior patterns (what used to be termed "instinctive behavior") that amount to fundamental mysteries as well. I think, for example, of the astounding abilities of migrating insects and birds to navigate to the right spot. Consider, for instance, the Monarch butterfly. These beautiful creatures travel southward in great swarms, and winter in specific places in California and Mexico; in spring and summer they migrate northward again, traveling to spots nearly two thousand miles away. However, they travel only part of the distance, lay eggs, and die. The next generation emerges from the chrysalis and continues the northward journey; and so on. Four generations after a butterfly leaves a eucalyptus tree near Santa Cruz, California, its descendants return *to the same tree!* This remarkable phenomenon seems at least a little easier to understand if one allows the possibility of a collective mind at the species level.

Many other migratory patterns pose similar questions. Sea turtles, after feeding for several years off the coast of Brazil, swim 2,500 miles to Ascension Island, which is only about eight miles across. Ascension Island is free of predators, so its appeal is apparent. But how do the turtles navigate to such a tiny speck in a vast ocean? They cannot be guided by the taste of island runoff in the sea, for example, because the arid land has practically no runoff.

Salmon find their way from hundreds and even thousands of miles offshore to the river and the same small stream from which they set out years earlier. Various navigational aids have been proposed, including the earth's magnetism, the position of the sun, the polarity of light, electric fields, and their detecting the taste of the native brook dissolved in thousands of cubic miles of sea water. Even with these possibilities in mind, it would seem impossible for salmon to do what millions of them do every year.

European eels are hatched in the Sargasso Sea not far from North America. Larval eels cross the Atlantic ocean to Europe, dawdle in the coastal waters, change shape, then head up rivers to mature. Eventually they head downstream and find their way back to the breeding grounds from which they set out years before.

SAHTOURIS:

Practically everything in biology, including these puzzles, relates to the central concept of evolution, neo-Darwinian or otherwise.

I'd like us to explore that area much more fully. In my book *Earthdance* (1996), I said:

Ever since Darwin our general view of evolution has been of a battle among individual creatures, pitted against one another in competition for inadequate food supplies. Only now are we in a position to understand the whole Earth as a living body—a single dance woven of many changing dancers and their complex patterns of interaction. Competition and cooperation can both be seen within and among species as together they improvise and evolve, unbalance and rebalance the dance. Evolution is this improvised dance in which ecological balance (mutual consistency) is worked out over and over.

Remember that living things have to change in order to stay the same; they have to renew themselves and to adjust to the changes around them. Rabbits evolve together with their "rhabitats," so to speak—all creatures evolving in connection with all else evolving around them.

It took a century and more after Darwin's theory was published for us to understand that environments are not ready-made places that force their inhabitants to adapt to them, but ecosystems created of, by, and for living beings . . . as they transform and recycle the materials of the Earth's crust. This view is also consistent with Ervin Laszlo's. (1996)

HARMAN:

Yes, evolution is such a fundamental metaphor in biology that almost all questions eventually get framed in its terms: How did these amazing instinctual patterns evolve? It's mysterious from the very beginning. Even the emergence of life appears very difficult to account for on a happenstance basis. Then, even at this level of the simplest form of life, Lynn Margulis identifies three major mechanisms of evolution that seem to have been important throughout this early phase of Earth's history: mutation of DNA, bacterial genetic transfer, and symbiotic alliance. Mutation appears to be not entirely random; there is evidence that at the single-cell level it can be quite purposeful. Genetic transfer also sometimes looks more than random. And we have already noted the phenomenon of "purposeful" symbiotic alliance.

SAHTOURIS:

Laboratory evidence, as I said before, comes in very strongly on the side of intelligent interaction between organisms and their environment at every level, including the genome, as soon as we open ourselves as scientists to seeing it. Let me quote my book again:

"Neo-Darwinism is a misleading way of seeing nature. The notion of the separateness of each creature, competing with others in its struggle against nature's challenge, fits society's own change to competitive and exploitative industrial production. Now, as we ourselves must learn to harmonize our ways with those of the rest of nature instead of exploiting it and one another ruthlessly, that notion makes little more sense than the notion that our own cells are separate beings competing with one another to survive in our hostile bodies. It is no longer useful or productive to see ourselves as forced to compete with one another to survive in a hostile society, surrounded by hostile nature.

We are feeling the pressure to change the whole framework of biology because we now see the extent to which it was formed to fit a political/economic climate and not to fit unbiased observations of nature. The first creatures of Earth, the bacteria that reigned supreme until two billion years ago, worked out an intelligent cooperative scheme of trading information and sharing out tasks that persists to this day. Not that they didn't go through phases of trying out competition, but that, too, was an informed process, and when it no longer worked they gave it up to cooperate as nucleated cells."

HARMAN:

That's a bold statement, that neo-Darwinism is a misleading way of seeing nature; yet I'm inclined to agree.

The basic concepts involved in Darwin's hypothesis regarding the origin of species were adaptation, inheritance (and variation), and natural selection. Organisms tend to be adapted to their environment. They are always in competition for the resources they need—food, safe nesting places, and so on. The variability among members of a population plays a central role; members possessing those characteristics that lead to better adaptation tend to survive. As the environment changes, the pressures of survival force corresponding changes in populations. This process may bring about the emergence of new, better adapted species and the extinction of those that fail to change adequately. The adequacy of this view (augmented by the detailed understanding of inheritance that came with the discovery of DNA) has been widely accepted.

However, if all of this is viewed with fresh eyes, there seems to be ample evidence to cast doubt on the completeness of the neo-Darwinist orthodox view of evolution. Perhaps the most general problem is the cumulative effect, over time, of organisms' self-forming *(autopoietic)* tendencies (see next chapter). Evolution itself has an autopoietic aspect not accounted for in the prevailing view. Then there is the amazing ability of organisms large and small to adapt to changing environments; the dialectical evolution of the unitary global system (Gaia); the relatively sudden appearance of new species, families, and phyla after the major extinctions; the incredible instinctual patterns throughout the animal kingdom—the list goes on and on. (See the "Interlude" following this chapter.)

For instance, there are many examples that suggest function can come before structure in evolution. Some of these were recognized by Darwin himself. I'm thinking, for instance, of the matter of *analogous* structures and functions. Take the eye, which seems to have been separately "invented" so many times. Numerous forms of eyes (mammal, insect, crustacean, etc.) have evolved, seemingly quite separately from one another and with distinctly different "solutions" to the prior function of seeing. And then there's flight. Biologists have made valiant efforts to guess how such a dramatic innovation as flight evolved in pterosaurs, birds, bats, and insects; but no scenario is convincing. Another such example is the mechanisms by which carnivorous plants (e.g. venus fly trap, pitcher plant) lure, capture, and digest insect prey. Were it not for the fear of appearing to have succumbed to the lure of anthropomorphism, one might be tempted to assume that organisms developed eyes because they "wanted" to see, that others developed wings to fly with, and that some plants "purposefully" evolved as insect-eaters.

Take such an obvious question as how the giraffe evolved. The acceptable answer is along the lines of mutants with longer necks, enabling them to reach higher leaves and hence tip the selection process in their favor.

This still leaves a lot for natural selection to explain. The protogiraffe had not only to lengthen neck vertebrae (fixed at seven in mammals) but to make many concurrent modifications: the head, difficult to sustain atop the long neck, became relatively smaller; the circulatory system had to develop pressure to send blood higher; valves were needed to prevent overpressure when the animal lowered its head to drink; big lungs were necessary to compensate for breathing through a tube ten feet long; many muscles, tendons, and bones had to be modified harmoniously; the forelegs were lengthened with corresponding restructuring of the frame; and many reflexes had to be reshaped. All these things had to be accomplished in step, and they must have been done rapidly because no [fossil] record has been found of most of the transition. That it could have all come about by synchronized random mutations strains the definition of random (Robert G. Wesson, 1991, p. 226)

Stephen Jay Gould (1989), however, offers the equally plausible suggestion that long necks proved advantageous in courtship rituals.

And then there is the issue of inheritance of acquired characteristics. As you implied earlier, Lamarckism has been widely assumed to be an erroneous concept. Molecular geneticists found it difficult to conceptualize how an animal or plant could respond genetically to external conditions. Yet many instances exist where it certainly looks as though organisms are reacting genetically to signals from the environment, including the examples you gave of organisms developing resistance to antibiotics, herbicides, and pesticides.

SAHTOURIS:

Scientists have, by and large, tended to deal with these kinds of examples one at a time and have sought to fit them into the "central dogma" of molecular biology, namely that there can be no transfer of information from organism to genome. But the evidence I cited earlier belies this dogma. To repeat, genes can be observed now as they are amplified, shuffled about, and traded among cells and even other organisms throughout development and even the whole life of the organism. Genes altered through interactions of the organism with its environment can be passed to its germ cells.

For me it is interesting to recall that in 1985, in the seclusion of a small Greek island where I was working at putting together a coherent story of evolution well before I knew anything about such researchers or their results, I wrote:

"It even seems ever more likely that DNA can reorganize itself—or be reorganized by other cell components—to repair the kind of accidental change that was thought to be the only way to evolution. It would be nice to think we are not just a lot of piled-up accidents and copying mistakes, but beings who have at least in part organized and evolved ourselves in harmony with the other living beings that form our environment."

Compare this with Mae-Wan Ho's quote at the beginning of this chapter, both written well before we knew anything of each other's work or ideas.

HARMAN:

Taking this comprehensive view, I am impressed with the implications of the remarkable "experimentation" with a multitude of "inventive" models that seem to characterize evolution. I know that there have been many attempts to explain the Cambrian explosion and similar episodes in the evolutionary story, but once you allow in something like consciousness, there is a strong allure to conceptualizing the periods of phenomenal radiation in terms of the metaphor of creativity.

When we think of evolution it is most often as what is termed "microevolution," comprising permanent, genetically based changes that occur within species as they differentiate into recognizably different races or subspecies living in different geographical areas or under different climatic conditions. Most biologists have little doubt that the chief Darwinian mechanisms of speciation—random mutation, geographical isolation, natural selection—are part of this picture. (We have already noted that mutations do not appear to be entirely random and may be quite "purposeful," however disturbing that thought may be to those who reject any thought of teleological influence in evolution.)

It is a different situation with macroevolution—the great changes in form, complexity of bodily organization, and mode of life that have occurred over the broad sweep of evolutionary history, as revealed by the fossil record and study of the comparative anatomy and physiology of extant organisms. The phenomenal Cambrian radiation preserved in the Burgess Shales (Gould, 1989) was not of species, but of phyla (that is, basically different body plans) and classes. (See the "Interlude" immediately following this chapter.) Following the mass extinctions about 550 million years ago, something of the order of a hundred new phyla appeared "suddenly" over a ten-million-year period, of which only about thirty survive today. These phyla do not appear to have resulted from a large number of species changes. Macroevolution awaits satisfactory theoretical explanation. It seems easier to conceptualize in the holarchic view than with the neo-Darwinist assumptions.

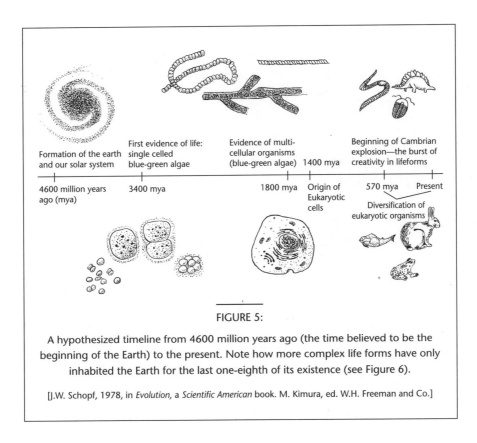

FIGURE 5:

A hypothesized timeline from 4600 million years ago (the time believed to be the beginning of the Earth) to the present. Note how more complex life forms have only inhabited the Earth for the last one-eighth of its existence (see Figure 6).

[J.W. Schopf, 1978, in *Evolution,* a *Scientific American* book. M. Kimura, ed. W.H. Freeman and Co.]

Radically new animal and plant forms appeared on the evolutionary scene with relative suddenness and without the expected transitional forms to bridge the enormous gaps that separate major divisions of organisms. The fundamental problem presented by the gaps in the fossil record is their systematic character.

It has been postulated that new species arise rapidly in peripherally isolated populations and then spread over a wide geographical area, undergoing little further change ("punctuated equilibrium"). This process may well explain gaps between species, but that still seems to leave unexplained the major discontinuities. For example, birds are rather generally assumed to

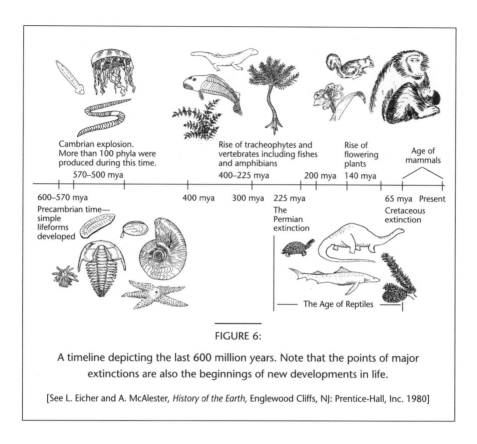

Cambrian explosion.
More than 100 phyla were
produced during this time.

570–500 mya

Rise of tracheophytes and
vertebrates including fishes
and amphibians

400–225 mya

Rise of
flowering
plants

200 mya 140 mya

Age of
mammals

600–570 mya

Precambrian time—
simple
lifeforms
developed

400 mya

300 mya

225 mya

The
Permian
extinction

65 mya Present

Cretaceous
extinction

——— The Age of Reptiles ———

FIGURE 6:

A timeline depicting the last 600 million years. Note that the points of major
extinctions are also the beginnings of new developments in life.

[See L. Eicher and A. McAlester, *History of the Earth,* Englewood Cliffs, NJ: Prentice-Hall, Inc. 1980]

have evolved from reptiles, but what kind of a reconstructed path
could one imagine by which the scales of the reptile evolve into
the feathered wings of a bird? Neo-Darwinism contains the
implicit assumption that all variations are relatively small, large
changes being made up of an accumulation of small ones. But
some changes are very difficult to conceive of this way—for
example, the mammal's limb becoming the wing of the bat
(appearance of the order Chiroptera).

As the distinguished biologist Ernst Mayr insists (1988), there
is "no clear evidence for any change of a species into a different
genus or for the gradual emergence of any evolutionary novelty."

The discoveries of molecular biologists make it clear that at a fundamental level of molecular structure, each member of a class seems equally representative of that class, and no species appear to be in any real sense "intermediate" between two classes. Nature, in sum, appears to be profoundly discontinuous. Neo-Darwinism can explain small gradual changes, but not the sudden appearance of dramatically new forms, which constitutes one of the major puzzles within this framework. It is a logical error to assume that if a certain amount of evolution can be accounted for by mutation and natural selection, the possibility follows that *any* degree of evolution can be accounted for in this way.

Doesn't that all look like a creative mind at work?

SAHTOURIS:

I imagine many biologists would balk at that thought, which I'll address in a moment. But first let me repeat that I see the whole neo-Darwinist picture of evolution as unsatisfying in the light of new data.

It seems that we have come to the time when it may be more useful to think of all your questions together and see whether they are not better answered within a new framework that begins from very different premises than neo-Darwinism—which, it seems to me, we are literally being forced to replace with something more adequate.

In particular, there are serious challenges to the chief Darwinian mechanisms of speciation (random mutation, geographical isolation, natural selection) as inadequate to account for the evolutionary data. As you say, there is much evidence suggesting that "mutations do not appear to be entirely random and may be quite 'purposeful,' however disturbing that thought may be to those who reject any thought of teleological influence in evolution." But I think the case is even stronger than that.

Brian Goodwin (1994b) points out that plethodontid sala-manders haven't changed their morphology in a hundred million years despite great changes in their habitats over that time and quite a lot of alteration in their genes. And Mae-Wan Ho con-firms this lack of relationship between genetic mutations and changing environments, which should cause the "natural selec-tion" of new forms, in observing that "variations are generated by the physiological interaction between organism and environ-ment, the persistence of the variations over the generations depends not on natural selection but on heredity! The variations would indeed persist in the absence of natural selection, so long as heredity operates in their favor. . . . There is a great deal of conceptual confusion here which some good philosopher of biol-ogy should try to sort out in future." (1988)

Just to break set and show that there are very different ways of looking at the whole process of evolution, I'd like to go into detail a bit here on the way the early 20th-century Russian geolo-gist V. I. Vernadsky understood the evolution of the Earth as a whole. Vernadsky's uncle, the Russian philosopher Y. M. Korolenko, is reported to have told him the Earth is a live being, though it is not at all clear that Vernadsky believed this. Nevertheless, his studies of Earth took a very different view of life than did those of other scientists: he called life "a disperse of rock," a geochemical process transforming magma into rock into creatures.

Vernadsky observed the Earth's crust turning cosmic radiation into its own forms of energy as it packages itself into cells and multicelled creatures, speeding its chemical changes with enzymes, transforming itself into ever-evolving creatures and back into rock. This view of living matter as continuous with, and as a chemical transformation of, nonliving planetary matter is very different from the view of life developing as individual creatures adapted to their environments. While this Vernadskian view stimulated much research in the Soviet Union, it never became widely known in the West, although it was introduced here by G. E. Hutchinson, Sterling Professor of Zoology at Yale University.

This concept of all living creatures together as living matter postulates that the living part of the Earth's crust is always energetic enough to actively transform the more passive parts into itself and its products. At first blush, this concept of living matter appears to be the same as Lovelock's (1988) concept of biota—the sum total of living creatures, contrasted with the abiotic, or non-living environment. But in Vernadsky's conception the emphasis is on geobiological continuity, on the cyclic transformation of the biological domain into the geological domain and vice versa, whereas in Lovelock's conception the emphasis is on their interaction as separate parts.

Oddly, Vernadsky, who apparently did not see the planet alive as a whole, perceived its integrity more fundamentally than does Lovelock. In the Vernadskian view one would see the same molecules as being part of rock in some epochs and part of living creatures in others. For example, he traced phosphorus through plants and animals to decayed matter and excrements, thence by erosion to the oceans, added to there by volcanic matter, some taken up in diatoms, forming sediments and fossils. Ocean floor sediment, containing vast quantities of algae and animal shells, is all passed through the guts of sand- and mud-eating worms to further transform it, just as soil is transformed by the related earth-eating earthworms of dry land.

The geological activity of creatures also includes their production of atmospheric gases and their transfer of groundwater back into the atmosphere, a process clearly visible in the pumping action of rain forests—the rain then falling to dissolve more earth and rock. On the whole, however, the geological activity of creatures is less the larger they are, most of this work being done by microbes and rock-, mud-, or earth-eating worms. Some microorganisms contain half a million to a million times as much of some mineral, such as iron, manganese, or silver, as their environment does, leaving veins of such minerals where they congregate and decay. Micro-organisms are even responsible for concentrating radioactive materials, such as uranium, possibly to keep themselves warm; the bower birds of Australia create chemical reactors as nests for their eggs.

Vernadsky thus understood metabolism as the activity of all living matter taken together as well as that of any particular organism. Since virtually all of the Earth's atmosphere, seas, soil, and rock, even its purest, hardest diamonds, are made from the dead bodies and by-products of organisms, it is clear that life is the most powerful of geological forces! The record of evolution lies in all of geology, not only in recognized fossils—as reflected in the title of a book on Vernadsky's work called *Traces of Bygone Biospheres* (Lapo, 1982).

An interesting notion of Vernadsky's was that biologists should reclassify living organisms on the basis of their individual, or species, metabolism. He argued that our present classification from kingdom to species by way of phylum, class, order, family, and genus had led us to classify as related many organisms that really are not related under natural conditions. A better scheme, he felt, would be to divide kingdoms according to the way in which each of their species metabolizes supplies from its environment, beginning with the autotrophs, including photo-autotrophs—"self-feeding" organisms that can build their own protein and nucleic acids from simple molecules and elements such as minerals, water, and carbon dioxide. The second major category would be the heterotrophs—organisms that feed off others because they cannot make large molecules from basic ones but must incorporate them by eating other organisms. The third category would be the saprotrophs—those that feed on the dead and thus reduce their large molecules back to the basic ones that autotrophs can use. The fourth category, mixotrophs, can metabolize in more than one way. Finer distinctions within these categories are made as heteroptrophs feed off other heterotrophs, and so on.

What is intriguing about this scheme is that organisms are classified not by their structure but by their functions within the whole geobiological life process. It recognizes organisms as self-organizing packets of the Earth's crust with enough energy to move about the more sluggish matter around them.

The energy of living matter sometimes explodes almost beyond belief. A locust plague of a single day has been estimated to fill six thousand cubic kilometers of space and weigh forty-five million tons! It is the locusts' heterotrophic metabolism, of course, that makes them a plague, as they suddenly convert vast quantities of the autotrophic crops, planted by humans, into their bodies, although these quickly revert to earth after their short lives. Most biogeologic activity goes on less dramatically—though it is impressive enough to consider that a single caterpillar may eat two hundred times its weight per day, and that earthworms pass nine to seventeen tons of earth per acre per year through their guts, transforming it into a rich medium for plants. Note that humans now destroy fertile land at a much higher rate than earthworms can create it.

Darwinian mechanics forced us to see individual organisms against a background environment to which they must adapt, a view that has become so prominent in our science that we can hardly even conceive of seeing, or ordering what we see, in any other way. I find Vernadsky a refreshing way to break this set and see from a new perspective, more consistent with my own view of evolution as a coherent improvisational dance.

I think I've made it abundantly clear by now that I am not a neo-Darwinist, that I see all nature as intelligent and interactive in this improvisational dance. I've been reluctant to call this "creative mind," because I have thought of it as intelligent in exactly the way our bodies are intelligent beyond the reach of our familiar conscious minds. While physiologists have long accepted the intelligence of the body, they do not endow it with "mind" in the usual sense, though I could certainly agree it displayed "mind" in Gregory Bateson's (1980) "mind of nature" sense. Your "metaphor of creativity" in evolution, on the other hand, seems to me an understatement. Where, if not in our observations of nature, did we first recognize creativity? Is natural creativity the metaphor, or is it the source of our metaphors for every human effort to be creative?

My metaphor for evolution as an improvisational dance seems ever more apt even at the genetic level, where each cell now appears to reassess, update and exchange parts of its DNA as it goes. The word "evolution," when used in talking about human dancing, means the changing patterns of steps in any particular dance. A dance thus evolves when its step patterns change into new ones as the dance goes on. In exactly this sense, the evolution of "Gaia's dance," of Earthlife, is the changing patterns of steps in the interwoven self-organization of creatures and their habitats over time. We see that this dance is endlessly inventive. Trying out new step patterns in a dance is called improvising, as a creative dance is not planned out in advance. Rather the dancers improvise as they go, testing each new step for its fit with other steps and with the whole dance pattern. Basic steps may be used over and over in new combinations, with new steps added at times. Like Brian Goodwin, I see this natural evolution as trying out whatever it can, without teleological intent, but constrained by the demands of mutual consistency.

In biological evolution, all creatures, from the first bacteria to ourselves, have been composed of DNA and protein molecules. The very complex patterns of these giant molecules are almost entirely made of only six kinds of atoms—hydrogen, carbon, nitrogen, oxygen, phosphorus, and sulfur. There are very few kinds of protein or other molecules on Earth today whose patterns the ancient bacteria had not already invented billions of years ago. Nor have any new basic life processes been developed since early prokaryotes invented the three ways of making the ATP energy molecules discussed earlier in the interlude on prokaryotes: fermentation, photosynthesis, and respiration. Vernadsky observed that organisms have incorporated and created 99.9 percent of all the kinds of molecules found in nature, almost all of them billions of years ago when bacteria were the only creatures around. In other words, evolution since then has been a matter of rearranging the same molecules and life processes into an endless variety of new creature patterns. This, then, is the Earthdance—the endless improvisation and elaboration of elegantly simple steps into the awesomely beautiful and complex being of which we are one of the newest parts.

Trying to maintain that the Earth can function in such ceaseless creativity and dynamic harmony for five billion years without transcendent self-reference (knowledge of itself) or conscious intelligence (self-aware information use) or even "creative mind" is to concoct an enigma, not to solve enigmas.

Interlude Two

Neo-Darwinism and Its Problems

The basic concepts involved in Darwin's hypothesis regarding the origin of species were adaptation, inheritance (and variation), and natural selection. Organisms tend to be adapted, often exquisitely so, to their environment. Offspring resemble their parents, but random variability occurs as well—differences in size, color, strength, speed, rate of growth, and so on. And organisms are always in competition for the resources they need—food, safe nesting places. Here the variability among members of a population plays a central role; members possessing those characteristics that lead to better adaptation tend to survive. As the environment changes, the pressures of survival force corresponding changes in populations. This process may bring about the emergence of new, better adapted species and the extinction of those that fail to change adequately. Thus the history of life is a tale of perpetual change in which history, heredity, and adaptation through competitive interaction are the ingredients of an evolutionary biology.

The basic concepts of genetics were discovered by Gregor Mendel (published in 1866 but ignored until the twentieth century); understanding of the mechanism involved came with discovery of the role of the double helix of DNA (deoxyribonucleic acid) by Francis Crick and James Watson in the 1950s. The synthesis of this mechanism of inheritance with the Darwinian concepts is what is generally termed neo-Darwinism. There should be no doubt that neo-Darwinism appears to explain a great deal. Were it not so, the theory would not have been as durable as it has proven to be, nor as universally adopted as the chief integrating framework for the biological sciences.

The neo-Darwinist view emphasizes random origins (by mutations); persistence of basic forms (via common inheritance); and successful functioning of the structures (adaptation through natural selection). Once organisms have hit upon a process that works, evolution tends to proceed through modifications of the fundamental form. An example often given is the similarity of limb structure among the tetrapods (four-legged vertebrates—amphibians, birds, reptiles, and mammals). Biological forms are thus the result of a combination of internal chance and external necessity.

There have had to be some modifications to the original theory as newly discovered information required. For example, as Richard Leakey has pointed out (1995), the role of natural catastrophes and multiple extinctions appears to be much greater, in comparison with natural selection, than was originally thought. He argues that there have been a half-dozen major extinctions throughout evolutionary history, some at least caused by devastating impacts from outer space in the form of comets or meteorites. During some of these extinctions the vast majority of all species living at the time vanished in a geological instant. One of these mass extinctions happened sixty-five million years ago, the most celebrated victims of which being the dinosaurs. (The most recent extinction is ongoing, caused by the devastating impacts of human societies.) However, by and large the theory has proven to be remarkably enduring and resilient.

Nevertheless, the evidence of a cumulative effect, over time, of organisms' self-organizing tendency in evolution; the remarkable ability of organisms large and small to adapt to changing environments; the dialectical evolution of the unitary global system (Gaia); the relatively sudden appearance of new species, families, and phyla after the major extinctions; the incredible instinctual patterns throughout the animal kingdom—all, if viewed with fresh eyes, seem to cast considerable doubt on the completeness of the neo-Darwinist orthodox view. In other words, the central question is whether the available evidence relating to evolution fits better within the strict neo-Darwinist picture, or is more comfortably accommodated within an expanded "holarchic" picture.

In the modern understanding of heredity, the basic characteristics of the organism are acquired through the DNA present in the zygote. In brief, this hereditary information contained within the DNA (the genotype) is essentially instructions for making RNA and proteins; these RNAs and their products are the chief building blocks for the organism (the phenotype). Their relationships, and their changes during embryonic development and evolution, are determined by the information contained in the DNA. The power of these concepts is amply demonstrated by the successful exploitation (in biotechnology) of such basic macromolecular processes as immunological (antigen-antibody) reactions, DNA replication (gene cloning), and DNA-RNA interactions (hybridization).

The neo-Darwinist interpretation of evolution builds on this understanding of the hereditary process. Genetic changes are to be understood as occurring basically through random mutations in the DNA. These changes in the genotype result in changes in the phenotype, where they increase or decrease its ability to adapt to its environment. Processes of natural selection, modified by geographical isolation, result in preservation or extinction of these new traits and, hence, of the aspects of the genetic material that carry them from one generation to another.

There is still some controversy over whether the interaction of the organism (phenotype) with its environment can affect the hereditary instructions (genotype). Jean-Baptiste Lamarck, whose work on comparative anatomy preceded Darwin's, insisted "yes," but August Weismann, in 1894, issued an authoritative "no." This latter answer was given additional weight by what was later learned about the mechanisms of heredity and became enshrined as what Crick called the "central dogma" of molecular biology, namely, that the sequence of information flow is from DNA to RNA to the proteins constructed by the RNA—and not the reverse flow. With this neo-Darwinist model, it is difficult to conceive of how acquired characteristics might be incorporated into the genotype; it contains the implication that changes in the genome could come about only by errors of replication. In other words, Lamarckism, the assumption that acquired characteristics can be inherited, seemed for a time to be denied by the neo-Darwinist conceptions. However, as we have seen, there is considerable evidence to the contrary.

Now let us contrast the neo-Darwinist picture with a holarchic one. In the latter there is no justification for insisting that purpose is to be found only at the level of human-holons and not below (as in "lower" animals) or above (as with Gaia). Nor is there justification for denying the possibility of something resembling consciousness, creativity, or experimentation in "less evolved" animals or at the Gaia level. That is not to "prove" that purpose and consciousness permeate the Whole; only to observe that in the holarchic view the possibility is not foreclosed.

What kinds of evidence might incline one toward this view? Let us look at a few examples.

1. One kind of evidence suggests that the remarkable instinctive behaviors and abilities of animals are such as to imply something like consciousness. Pet owners and naturalists are typically convinced of this simply by their observations. One can find many examples in the wild. Consider for example the well-known clever ways of nesting birds. Many species of weaverbirds (family *Ploceidae*) construct elaborate nests of interlaced vegetation. Some tie strands of grass in overhand knots, not an easy task without hands. Some braid a sort of rope to hang the nest as much as a meter below the supporting branch. The Indian tailorbird (family *Sylviidae*) sews broad leaves together with strands of fiber to support and hide their nests. To simply call these behaviors "instincts" or "innate behavior patterns," and to argue that they are due to some not-yet-discovered "program" in the DNA, explains nothing.

The European cuckoo poses a similarly puzzling question. Due to the cuckoo's unneighborly habit of laying their eggs in other birds' nests, the young are hatched and reared by birds of other species, and never see their parents. Towards the end of summer, the adult cuckoos migrate to their winter habitat in Southern Africa. About a month later, the young cuckoos congregate together and then they also migrate to the appropriate region of Africa, where they join their elders. They instinctively know that they should migrate as well as when to migrate; they instinctively recognize other young cuckoos and congregate together; and they instinctively know in which direction they should fly and where their destination is. As Sheldrake (1981) argues, this is among a wealth of evidence suggesting something like a "morphic field" of species-level consciousness.

2. An important aspect of this evidence is the behaviors and abilities that suggest a collective consciousness beyond the individual organism. This is well known in the behaviors of schools of fish or flocks of birds, where vast numbers of organisms seem to change direction at the same instant. It is tempting to explain this phenomenon away by assuming some sort of visual signal. However, there are many other examples not so easily disposed of.

Consider, for instance, the lowly termites. These ancient social insects create elaborate mounds that can be as much as twenty feet high, being home to several million insects. Some of the chambers of these mounds may extend far into the ground, with networks of subterranean passages and overground tubes leading into the surrounding area where workers collect food. Typically, the thick, hard outer wall of the mound contains air ports and ventilation shafts. Within the nest are many chambers, passages, and the fungus gardens within which the termites cultivate fungi on finely chewed wood. These structures are built by hundreds of thousands of workers from pellets of soil first moistened by saliva or excrement.

In the making and repairing of nests, the termite workers do not simply respond to each other, but also to the physical structures that are already in place. For example, in making arches, workers first make columns and then bend them towards each other until the two ends meet. The workers on one column cannot see the other; they are blind. They seem to "know" what kind of structure is required, as though responding to some kind of "plan."

The South African naturalist Eugène Marais made a series of observations of the way workers of a *Eutermes* species repaired large breaches he made in their mounds (reported in Sheldrake, 1994). The workers started repairing the breach from every side, each carrying a grain of earth which it coated with its sticky saliva and glued into place. The workers on different sides of the breach did not come into contact with each other and could not see each other, being blind. Nevertheless, the structures built out from the different sides joined together correctly. The repair activity seemed to be coordinated by some overall organizing field.

Marais experimented by driving through the center of the breach a steel plate, a few feet wider and higher than the termite mound. Thus the wound, and the mound, were divided into two separate parts. The builders on one side of the breach could presumably know nothing of those on the other side. In spite of this, the termites built a similar structure on each side of the plate, and when the plate was then withdrawn, the two halves matched perfectly after the dividing cut had been repaired. In another experiment, the steel plate was driven in first, and then a breach made on either side; again the termites built matching structures on either side. Mysteriously, if the experiment was repeated and the queen of the mound was destroyed or removed, the whole community immediately ceased work on both sides of the plate.

3. Another puzzle involves abilities that seem to imply something beyond our usual understandings of brain and physical mechanism—e.g., migration. Birds, turtles, fish, and sundry other animals find their way over vast distances. Birds fly without previous experience as far as from the Arctic to the Antarctic, often without landmarks. The three-gram ruby-throated hummingbird goes 600 miles nonstop across the Gulf of Mexico. The young of the bronzed cuckoo, a month after being left behind by their parents in New Zealand, fly 1,200 miles over water to Australia and then 1,000 miles north to the Solomon and Bismarck Islands.

It has been postulated that birds navigate by the stars or by the Earth's magnetic field. Yet no one has any idea how the fliers hit small targets after long journeys, often in the dark, during which they may have been blown off course by winds. They keep on course even when clouds hide the stars above and the Earth below.

Not only strong fliers like the plover, but also snipes, curlew, and sandpipers, migrate between Alaska or Siberia and the Hawaiian Islands; missing the islands in the featureless ocean would be fatal for nonoceanic birds. The Hawaiian Islands were never connected with the mainland, and they are only a few million years old; it is not easy to imagine how the instinct to migrate seasonally 2,500 to 4,000 miles from Alaska or Siberia could have started.

The navigational ability of pigeons has been studied intensively for many years. There is some evidence that they may orient themselves by the Earth's magnetic field and the position of the sun. But their faculty for knowing the direction of home from a distant region where they have never been before still defies understanding.

4. Another body of data suggests "directed" evolution, inheritance of acquired traits, and a reassessment of the Lamarckian hypothesis. We considered this controversy above. The discovery in the 1950s that nucleic acid (DNA and RNA) is the carrier of heredity, demonstrating the physical materiality of the genes, comprised a major step in understanding inheritance. Research emphasis shifted from the organism to its reproductive material. The organism is considered as a bundle of traits, each associated with one gene or a combination of genes. Genes are pleiotropic (have multiple effects); thus useless traits can be considered to have come along with useful ones. Major changes can be understood in terms of regulatory genes, which turn on or off "batteries" of structural genes. Less easy to explain are traits that are not advantageous for the individual but are so for the group.

With the neo-Darwinist model, as mentioned above, it is difficult to conceive of how acquired characteristics might be incorporated into the genotype. The "central dogma" of genetic biology insists that an animal or plant cannot respond genetically to external conditions. However, many examples exist where it certainly looks like organisms are reacting genetically to signals from the environment. Hundreds of species of insects and mites have developed strains that are resistant to the pesticides that were introduced to control them, sometimes within a season or two. All or almost all insects that humans try to kill with pesticides fight back genetically. The potato beetle in two years acquired immunity to nine pesticides. Of course all this could be explained on the basis of multifold accidental mutations, one of which was the right one, and natural selection. But the amount of straining to believe such an explanation is an indication of the strength of the "central dogma."

In wasps, bees, and ants the difference between queens and workers is conditioned by nutrition, pheromones, and/or behavior; their genetic material is the same. Very little is known about how plants (seedling, cutting) register environmental conditions and direct cells to activate the necessary genes to become stem, roots, leaves, flowers—all these diverse cells having the same genetic makeup.

"The impossibility of input from environment to the genome has never been proved. In view of the inventiveness of life, it would be surprising if a law of nature absolutely prohibited it. . . . A camel is born with callosities on its knees; can this hereditary trait be detached from the ability of the skin to thicken in response to friction? . . . Insects, crustaceans, salamanders, and so forth lose eyes in caves, in some cases rather quickly. . . . If bacteria are placed in a medium with nutrient molecules too large to pass through the pores of its membrane, they mutate to make the membrane more permeable. . . . The question is not whether organisms can respond genetically to external signals but the range and kind of signals to which they can respond. Morphogenesis is a continuous process of feedback between cells and their environment in the broadest sense, in which the whole system exerts control, and adaptation is not merely of genes but of the entirety." (Wesson, 1991, pp. 227–240)

5. There are indications that the DNA explanation of heredity is not sufficient. For example, one often finds statements to the effect that mutant genes "cause" particular types of change of form or morphology in organisms. An example is a homoeotic mutant called antennapaedia in the fruit fly *Drosophila,* in which legs appear during the embryonic development of the fly where antennae would ordinarily arise. However, this is cause in neither a specific nor a sufficient sense. It is not specific, because the effect of the mutant gene can be produced in normal (non-mutant) flies by a nonspecific stimulus, such as a transient change in the temperature to which the embryo is exposed at a particular time in its

development; and it is not sufficient, because knowledge of the presence of the mutant gene is not enough to explain why the morphology changes as it does.

6. One of the arguments put forth supporting the Darwinian concepts is that of homologous structures—similar structures in different species, often serving different ends. The prime example of a homologous structure is the pentadactyl limb in mammals (e.g., the hand of a human, the fin of a whale, the leg of a horse, the wing of a bat, and the paddle of a seal). The case for Darwinism is weakened by the facts that (a) homologous structures are not found to be specified by homologous genes, and (b) they are not found to follow homologous patterns of embryological development.

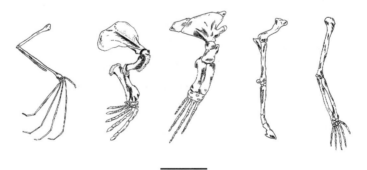

FIGURE 7:

Example of homologous structures. The right pentadactyl limb of five different mammals: the wing of a bat, the paddle of a seal, the fin of a blue whale, the leg of a horse, and the arm of a human.

With regard to the latter:

In some ways the egg cell, blastula and gastrula stages in the different vertebrate classes are so different that, were it not for the close resemblance in the basic body plan of all adult vertebrates, it seems unlikely that they would have been classed as belonging to the same phylum. There is no question that, because of the great dissimilarity of the early stages of embryogenesis in the different vertebrate classes, organs and structures considered homologous in adult vertebrates cannot be traced back to homologous cells or regions in the earliest stages of embryogenesis.

Like so much of the other circumstantial 'evidence' for evolution, that drawn from homology is not convincing because it contains too many anomalies, too many counter-instances, far too many phenomena which simply do not fit easily into the orthodox picture. . . . It is true that both genuine homologous resemblance . . . and the hierarchic patterns of class relationships are suggestive of some kind of theory of descent. But neither tell us anything about how the descent or evolution might have occurred, as to whether the process was gradual or sudden, or as to whether the causal mechanism was Darwinian, Lamarckian, vitalistic or even creationist. (Michael Denton, 1985)

7. Many examples suggest that function can precede structure in evolution. The diversity of forms taken by organs of

vision that presumably evolved separately from one another (eyes of insects, reptiles, crustaceans, mammals, etc.) is a commonly cited example. One can no doubt imagine these markedly different arrangements (*analogous* structures) for achieving the common "goal" of vision to have each come into being through random mutations and then persisted because of the advantage they bestow. But when one considers the complexity of each, and the number of simultaneous changes required for a functioning eye to appear in a previously sightless creature, this sort of accidental evolution seems implausible indeed.

8. Evolution achieves many things apparently beyond the powers of natural selection. For example, there are many organs and innate behavior patterns for which it is difficult to envision viable intermediate stages or which require different unlikely changes to come together for utility. Besides the examples already mentioned, another is the electric apparatus of fish. Some fish, combining the abilities to generate electrical impulses and to perceive them, use an electric field up to several volts as a sort of radar, sensing disturbances caused in the field by other fish or solid objects. The electro-sensing fish generally live in turbid water or are nocturnal, and have weak eyes.

Other electric fish use their discharge as a weapon. Electric eels *(Electrophorus),* with some six thousand generating plaques, can produce about one ampere at 500 volts; they do well not to electrocute themselves.

For electrolocation, many things must work together: an apparatus to generate fairly strong electric pulses at a rate of as many as seventeen hundred per second, consisting of a large number of plates stacked up in series; effective insulation of the electric generator from the body to make it possible to pile up voltage without allowing it to leak backward; special fins to swim without flexing the body and thus disturbing the field; a means of controlling the pulses; extremely sensitive receptors capable of registering minute changes in the strong primary gradient of the field; means of filtering out the electric discharges of other fish, which are much stronger than the faint echoes of its own field; and a special structure in the brain to process and use the information gathered.

Robert G. Wesson (1991) describes, among many examples, the human botfly's way of laying its egg on a person's skin. It waits until a mosquito or a bloodsucking fly approaches, then catches it and sticks its egg to the mosquito's or fly's belly. When the mosquito alights on warm flesh, the botfly larva emerges rapidly from the prehatched egg, drops off, and burrows into its new host. The botfly must have developed an innate behavior pattern to pursue and catch the mosquito or fly, grasp it and glue its egg to the underside; the egg must respond to the warmth of the mammal (only slightly higher in temperature than the air); the larva must release itself, drop onto the human flesh, and bore in. No portion of this complex pattern is very useful by itself.

In the big picture, revolutionary change seems to have been caused much less by environmental pressures than by the ability of organisms to find new means of making a living—means that probably do not arise from external conditions but from the genetic structure and the shifts of which it may be capable. Land animals became whales millions of years after the demise of large marine reptiles, not because the ocean became more inviting but because the pre-whale became capable of that metamorphosis.

9. Something like creative imagination is suggested by many aspects of evolution. Take the spider's web, for example. To explain this on the basis of mutation would require the cooperation of a large number of genes, gene families, or combinations of families. The web requires cells to produce different kinds of silk for different purposes (strong cables, sticky nets, wrapping), means of storage, channels and a pump for its extrusion, plus adaptations for handling the threads and a large set of instincts to construct the web and make good use of it—involving no doubt thousands of genes, most of which are purposeful only in the context of the way of life that they make plausible. If, as seems likely, the whole could not have been put together by random mutations bringing each individual gene into existence, modules comprising many genes must have been brought together, probably by sexual recombination. This is a very different task for natural

selection than a simple shift of gene frequencies, as postulated by population genetics. Natural selection must mostly be not for particular traits but for apparently unrelated families of traits.

10. One of the recognized puzzles in evolutionary theory is that posed by macroevolution—the great changes in form, complexity of bodily organization, and mode of life that have occurred over the broad sweep of evolutionary history, as revealed by the fossil record and study of the comparative anatomy and physiology of extant organisms.

Microevolution comprises permanent, genetically based changes that occur within species as they differentiate into recognizably different races or subspecies living in different geographical areas or under different climatic conditions. There seems little doubt that the chief Darwinian mechanisms of speciation—random mutation, geographical isolation, natural selection—are part of the picture. As noted above, however, there is considerable evidence suggesting that mutations are not entirely random and may entail a "purposeful" component.

The macroevolution of animals appears to be quite a different story. The animal kingdom (metazoa) is divided into major subdivisions called phyla; each phylum represents a basically different body plan. (For example, the phylum Chordata contains all the vertebrates, including as classes mammals, reptiles, birds, amphibians, and fishes. Other phyla include the Arthropoda [including insects, arachnids, and crustaceans], Cnidaria [including corals and jellyfish], and Mollusca [mollusks].) It does not appear that the phyla resulted from a large number of species changes.

The big branching of metazoan life began very early. The Cambrian radiation of new life forms, following the earliest known mass extinction around 550 million years ago, was a period of unprecedented evolutionary "experimentation." Something of the order of a hundred new phyla appeared "suddenly," over a ten-million-year period, of which only about thirty survive today. On the other hand, there are far more species today than half a billion years ago.

Over time the phyla, classes, orders, families, and genera have settled down. In recent times evolution has seemed able to generate only species. No new phyla have appeared since the early Cambrian period; no new classes for at least 150 million years; no new orders since the post-dinosaurian radiations some sixty million years ago. The basic radiation of mammals lies fifty million years or more in the past.

CHAPTER
THREE

Autopoiesis and Holarchies

Science is an attempt to bring the chaotic diversity of our sensory experience in correspondence with a certain unified system of thinking.

—Albert Einstein

But what is life? And matter—which is in perpetual continuous and lawful motion, where endless destruction and creation occurs, where there is no rest—is it dead? Can it be that only the hardly noticeable film on an infinitely small dot in the universe—on the Earth—possesses radical, specific properties, and that death is reigning all over beyond it? For the time being one can only bring up these questions. Their solution is sooner or later given by science.

—Vladimir Vernadsky, (1884)

SAHTOURIS:

In discussing these enigmas of biology and alternative ways to see them, we've prepared the ground to lay a careful foundation for our new model, which I'd like to do now. The first two of four basic features of this proposed model, as presented in Chapter One, were first, autopoiesis or self-creation as the fundamental definition of life, and second, the understanding of living systems as embedded within other living systems holarchically all the way from microcosm to macrocosm. We discussed holarchy at some length in that chapter without bringing in the basic concept of autopoiesis, which I'd like to do now.

Biologists have observed that living systems, from the smallest microbes to the largest organisms, exhibit self-organization, though it has never been explained by physical principles. That living systems exhibit this lifelong self-organization and preservation of integrity is evident, for example, in their ontogenesis from a single fertilized egg or seed to an adult; in their homeostatic and homeorhetic physiology; in their healing or regeneration; and in their complex innate behavior patterns for protection and reproduction. Thus, all of life is basically defined by this self-generating, self-maintaining criterion, as the Chilean biologists Humberto Maturana and Francisco Varela (1987), later at M.I.T. and the University of Paris, respectively, acknowledged a few decades ago. Accordingly, they proposed *autopoiesis,* a Greek word meaning 'self-formation' or 'self-creation' as the core definition of life. By this definition, an autopoietic (living) entity is one that continually creates its own parts, including its boundaries.

Before we had this definition, there was as James Lovelock pointed out no fundamental definition of life anywhere in the sciences. Each science seemed to think some other branch had one. Biology teachers and professors could only list observable properties of living things that distinguished them from nonliving things, the list including such properties as irritability, motility, growth and reproduction. Now that we have autopoiesis as a definition, we can see that some of these properties, such as growth and reproduction, are not part of the essential life process and thus may or may not be properties of a living system. Individuals (holons) do not have to reproduce to be alive.

Suppose we could show that it makes sense to order our universe as holons in holarchies, and to see each holon as alive by this autopoiesis definition of life. Then the whole universe would be seen as self-creating, self-evolving living systems at every level—cosmologic, astronomic, planetary, cellular, and perhaps even molecular and atomic. This is the model I have adopted—the universe as an autopoietic, self-creating living system—much as did Erich Jantsch (1980), who drew heavily on Ilya Prigogine's work.

Let me try making the case for this conceptualization. For if such a model could be shown to be both intuitively and intellectually more satisfying than the present model of a nonliving universe in which life happens by rare and still mysterious accident, then it would also become reasonable to propose that biology, as the study of living things, should become the fundamental science. We could also then envision that this unified holarchic worldview of a living universe would ultimately be shared by biology and physics, with much cross-pollination among their various specialized disciplines.

HARMAN:

Let's take this on a bit more slowly. As I understand what you're saying, autopoiesis is a fundamental mystery. That is to say, in the worldview of Western science it simply doesn't fit. As I think about this, it seems difficult to believe that autopoiesis just happened by chance encounter, and then autopoietic beings prevailed through natural selection. Autopoiesis is too pervasive and too efficacious; I believe we must examine Western science to see why it doesn't fit.

You want to go much further, and extend this concept of the living universe out to large and small extremes where it is not generally considered to apply. I'm willing, but let me try to ground the discussion first by considering a particular life-form.

It may be helpful to think about the very simplest life-forms we know, and the earliest in the conventional sense—the prokaryotes (one-celled organisms that lack a nucleus). These simple organisms do indeed seem to be self-forming, self-sustaining, and self-modifying. The energy for the processes involved is obtained through metabolism of various substances obtained from the entity's environment. The continuity throughout evolution comprises an unbroken line of these metabolic processes that began over 3.5 billion years ago.

Even in the very simplest cell to be found today, there are a number of essential components:

Membrane. Required for the maintenance of an internal cellular environment. Made up of lipids, saturated and mono-unsaturated fatty acids, glycenol, and proteins.

Cytoplasm. Fills the body of the cell; necessary for transport and as a solvent.

Enzymes. For catalysis, that is, speeding up chemical reactions. Involves proteins, peptides, amino acids, and nucleotide coenzymes.

RNA. For protein synthesis, transfer, messenger function.

DNA. Necessary for replication.

That's quite a bit to imagine coming together "accidentally," but in the prevailing cosmology it must have.

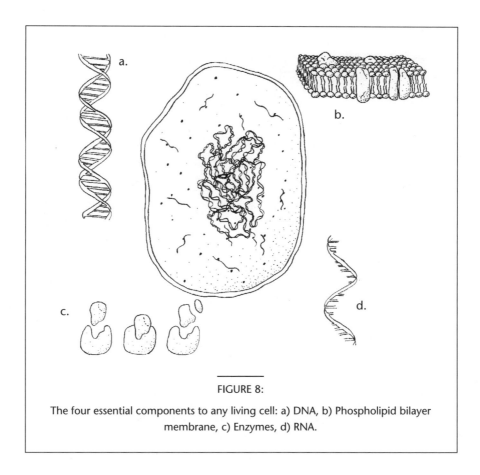

FIGURE 8:

The four essential components to any living cell: a) DNA, b) Phospholipid bilayer membrane, c) Enzymes, d) RNA.

In this "minimum cell" the essential processes—capture of energy and intake of matter at the cell's surface, production and repair of structural components, extrusion of energetic and material wastes, and so on—can now be described by scientists in considerable detail, but a mystery remains regarding their underlying dynamic; it was appreciation of this that led earlier investigators and philosophers to postulate a "life force," an *elan vital*. This idea—vitalism—was drummed out of the biological sciences generations ago, but it, or something like it, may be coming back.

What seems to be the most generally accepted theory of the origin of life proposes that chemical reactions on the early Earth gave rise to the essential building blocks of life (e.g., purine and pyramidine bases, amino acids and sugars), which then combined to form nucleotides, primitive nucleic acids, and primitive proteins. For life to arise in this "primeval soup" requires the appearance of a molecule that can replicate, that can make new generations of molecules like itself. The only "biological" molecules that can do this are the nucleic acids RNA and DNA, through which present-day organisms pass on genetic information from generation to generation. To get from the properties of RNA (which is generally assumed to have developed before DNA) to the minimal criteria for life by pure "accident" strains the imagination, yet according to this theory it must have happened.

It is commonly supposed that RNA must have arisen very close to the origin of life, because it can uniquely act both as self-reproducer and as catalyst. But RNA is difficult to make; it is difficult to conceive of its coming into existence by a chance combination; and unless there is a guidance mechanism it does not reproduce itself exactly. At a very early stage of life, there had to be a set of protein structures to permit nucleic acid to function, yet nucleic acid was necessary to make the proteins. There had to be some sort of membrane to contain interacting proteins and nucleic acid, but proteins and nucleic acid were necessary to make the semipermeable membrane to admit useful materials and permit waste to diffuse out. It is hard to see how life could possibly have put itself together bit by bit.

Now I believe that what you are saying is that this all becomes more understandable if we think in terms of a universe that is alive all along, before the first so-called "living organism" appeared on the scene.

SAHTOURIS:

Exactly. You are quite right that autopoiesis appears mysterious in the context of the Western scientific worldview, because autopoiesis defines life, and life itself is very mysterious in that worldview, since it must literally, as you say, "put itself together bit by bit." These "bits" are assumed to come from nonliving matter, so life is held to come from nonlife, which I find no more credible than the idea that intelligence comes from nonintelligence or consciousness from nonconsciousness. More on all that later.

The work of Prigogine and Jantsch (1983) probably goes as far as one can go in making autopoiesis "fit" the prevailing scientific paradigm. Prigogine's autopoietic dynamics of self-organization in simple chemical dissipative structures far from equilibrium and Jantsch's descriptions of Manfred Eigen's simple molecular system "hypercycles" (closed circles of transformatory or catalytic processes in which one or more participants act as autocatalysts) go far to reduce the gap between the animate and the inanimate in the prevailing view. They begin with the minimal molecular system and show that it can work to create and maintain itself. This effort to close the gap between life and nonlife necessarily takes the process of self-creation down to very simple chemical levels; yet, an interesting consequence of this work is that such simplification opens up the possibility of closing the same gap at macroscopic levels of the early universe, when things were also relatively simple.

Indeed, this is the way I myself, after Jantsch, conceptualized the living universe: that autopoiesis happens from the beginning at all size levels, from macrocosm to microcosm. I will describe this process shortly. Yet I want to say now that I have come to see this as fooling ourselves into believing that we had *explained* these simple autopoietic systems by *describing* their evolution. Now I am persuaded that we will have to go beyond squeezing life into the accepted models of "nonliving" physics and chemistry. That is exactly what has stifled biology for so long, as you yourself say, much as one has to admire the heroic efforts to make it fit.

Autopoietic entities in my view do not appear miraculously out of the chemistry of a nonliving universe, nor do they evolve by natural selection, which is really also just a descriptive and not an explanatory concept. Darwinian and neo-Darwinian evolution, as Lynn Margulis emphasizes, tells us something about survival but almost nothing about what really drives evolution, except for the assumed random mechanical copying error or damage to genetic material as the chief or only source of change. As I believe such damage is repaired—and we hear now of "repair genes"—rather than becoming the source of variation, I cannot subscribe to the Darwinian hypothesis on this matter. As for the tautological concept of natural selection or adaptation, I prefer the concept of active interspecies negotiations, which I call the process of working out mutual consistency. My view of Darwinism is that it is an attempt to fit life and its evolution into nonliving mechanistic models that include randomness generators as part of the machinery, along with a kind of cogs-and-wheels view of environment and species.

Clearly we need something much more powerful to explain evolution. Arthur Koestler commented that life is far too intelligent to evolve by accident, and it seems more and more biologists are coming to agree with him. As long as we exclude intelligence and consciousness, both cosmic and planetary, from our story of evolution, we will have description without explanation. By the time we have finished this discussion, I hope we will have shown persuasively that intelligence and consciousness are included in our model *without any dualistic assumptions about mind and matter,* including those of some life force inhabiting matter. But I think we need to continue with the description of autopoiesis—the *how* of life at all levels—before we attempt to explain *why* it happens, which will involve consciousness.

As I began using the autopoietic definition of life, it seemed to me that it fit not only single-celled creatures, fungi, plants and animals—all those categories Western science calls "kingdoms of life"—but also the Earth as a whole. Let me discuss several aspects of this.

(1) Autopoiesis: the living Earth. Biologists have been working with the assumption that Earth is a nonliving geological ball upon the surface of which, by some miracle, life sprang from nonlife. Yet we know that the Earth constantly creates itself from the inside out, lava erupting from its molten insides to form new rock, while old rock erodes, is incorporated into microbes and organisms, is carried to the oceans as both "inorganic" and "organic" matter, is settled into sediments, and is eventually remelted at the subduction zones of tectonic plates. All this constitutes one great crustal recycling system; other recycling systems involve Earth's waters, soils, and atmosphere. Together these cycles are sources of endless creativity, endless supplies of materials to be incorporated into evolving microbes, plants, and animals. Recall Vernadsky's view of these cyclic geobiological transformations.

If we accept autopoiesis as the definition of life, then we can clearly see that the Earth is alive. Anyone arguing that these are simply nonliving processes has to account for the fact that this entity in its entirety has now evolved increasing complexity for

five billion years with no signs of moving toward entropic disso-
lution or crystallization! That its life will end and its materials be
recycled in another estimated five billion years, due to the imminent death of its star, the Sun, is another matter—a matter of life
and death in a living universe.

When I first proposed to Francisco Varela that by his autopoietic definition the Earth was alive, he responded that my detailed
arguments about how the Earth continually creates and recreates
itself were persuasive, but he objected that the boundary of Earth
(its atmosphere) was too fuzzy to qualify as a proper boundary. I
suggested he compare a cell blown up to the size of a planet to
see which had the fuzzier boundary. Even in a photograph of
Earth from space, we can see how nonfuzzy its atmospheric
boundary appears at that size.

Our human size and time perspective often make it difficult
for us to see the relationships and processes of life at larger, slower scales. If only we could see the Earth's life as a twenty-minute
film, with all its eruptions and gyrations of continental drift, its
changing colors of evolving seas, land, and atmosphere, the
diaphanous swirl of its cloud skin, we would not doubt it was
alive any more than does Lewis Thomas in *Lives of a Cell* (1975),
where he calls it a great cellular creature "marvelously skilled in
handling the sun." I often see the elegance of nature in the
ancient dictum "As above, so below," and so find appealing this
metaphor of the Earth as a giant cell within whose boundary
membrane other smaller cells multiply, die, and are recycled in
such a way that the whole need not grow. This is a wonderfully
efficient way to make living beings (planets) possible with only
star energy for nourishment.

In 1785, the Scottish scientist James Hutton, remembered as
the father of geology, called the Earth a living superorganism and

proposed that its proper study should be physiology. He seems to have been virtually ignored, until the English atmospheric scientist James Lovelock took up this idea in the early 1970s with his Gaia Hypothesis (*Gaia*, in the modern form *Gi*, remains the Greek name/word for Earth). Lovelock has demonstrated the Earth's metabolic self-regulation by way of mutual interaction between its living organisms and their influence on environments.

Note that the Earth can function as a living entity only because all its parts are in constant interaction and because of the ceaseless planetwide flow of its energy and materials. Life, as Teilhard de Chardin, a close associate of Vernadsky, stated long ago (1959), and as Lovelock confirmed in the present, will never evolve naturally on only one part of a planet but instead will quickly spread around its sphere. Now we can see that entire planets, including their "insides," either come to life as wholes or do not come to life at all.

Why did Earthlife become so diversified as it evolved? We might as well ask why the first gas clouds sorted themselves into individual galaxies, and the galaxies into stars and planets and other space bodies. The answer, as we now begin to understand, may be that life becomes ever more stable as it becomes more complex!

Mechanical systems may be more vulnerable to breakdown as they become more complex, but this seems not to be true of living systems. The Gaian division of labor or function among different species—different kinds of creatures—makes possible a division of labor similar to that of our bodies, which function efficiently through the combined work of many different kinds of organs. No place, or environment, on Earth—from the barest mountaintop to the deepest part of the sea—has fewer than a thousand different life species or forms, mostly microbial, living lifestyles that keep it alive and evolving. Again we see that if a planet *does* come alive, it would seem that it must come alive everywhere, not just in patches.

To repeat the prokaryote story just a bit in this context, the giant molecules from which the first creatures formed themselves were apparently catalyzed by powerful solar and lightning energy, and perhaps by energy from Earth's hot core as well. Some of this energy was incorporated into the first fermenting prokaryotes, which transformed and released it in using it to break up other big molecules for food. Photosynthesizing bacteria evolved in learning to perpetuate themselves by using solar energy directly, maintaining themselves, and eventually producing the oxygen-rich atmosphere in their process. Oxygen-burning respirers got their energy by consuming the ready-made large molecules of fermenters, photosynthesizers and one another. Organisms can thus convert stored energy or solar energy into other useful forms of energy—the energy of motion, of heat, of chemical reaction, even of electricity.

These early autopoietic bacteria had enclosed their metabolic molecular systems within open boundaries—membranes of their own making that not only allowed materials and energy to be exchanged with the environment but also permitted them to exchange genetic material with each other, as all bacteria can do to this day. Margulis and Dorion Sagan have said that bacteria trade genes "more frantically than a pit full of commodity traders on the floor of the Chicago Mercantile Exchange" (1995). Margulis earlier pointed out that this is original sex, which had nothing to do with reproduction (1986). This makes it impossible to speciate bacteria, but it permitted them to function as a single system, a superb planetary skin. They traded bits of DNA endlessly with each other across those membranes, diversifying their lifestyles, living off each others' wastes, trading other materials with each other and with the crust, evolving new ways of making a living under environmental challenges, stabilizing Earth's chemistry and energy economy, creating an oxygen-rich atmosphere and a protective ozone shield, all in the course of a few billion years, about half the life of Earth itself. To this day these marvelous unseen creatures provide the basis for our own lives.

When at last, around a billion and a half years ago, they abandoned competitive lifestyles and assembled themselves into cooperative communities, they formed the only other type of cell ever to evolve: the nucleated cell, or eukaryote (on the average a thousand times larger than a prokaryote). Each bacterial member of these large communities gave up some of its DNA, along with its independence, to form the "central library" of genes we know as the nucleus. In a division of labor, aerobic bacteria became mitochondria, blue-green bacteria became chloroplasts, and so on.

To this day, bacteria and the monera (nucleated cells) they formed remain the fundamental creatures of Earthlife. We ourselves as individuals, along with all other animals and plants, are cloned at every generation upon versions of the ancient single-cell cooperatives of these once free-living bacteria.

Lovelock's first clue to Gaia came to him when he was comparing the atmospheres of different planets. The atmospheres of the other planets in our solar system all make sense chemically— they are stable mixtures of gases. Only Earth has an atmosphere that is quite impossible by the laws of chemistry. Its gases should have burned each other up long ago! Yet if they had, Earth would have no living creatures. As it is, every molecule of air we breathe has actually been recently produced or recycled inside other living creatures. Earth's creatures make and use almost the entire mixture of gases in the atmosphere, with the exception of trace amounts of inert gases such as argon and krypton, ever feeding the mixture new supplies as they use it and as it burns itself up chemically.

This activity of living things always keeps the atmosphere in just the right balance for the life of Earth to continue. Living creatures, for example, produce four billion tons of new oxygen every year to make up for use and loss. They also make huge amounts of methane, which regulates the amount of oxygen in the air at any time, and they keep the air well diluted with harmless nitrogen. In fact, the atmosphere is held at very nearly 21 percent oxygen all the time. A little more and fires would start all over our planet, even in wet grass. A little less and we, along with all other air-breathing creatures, would die.

If organisms, especially micro-organisms within the great Gaian system, stopped making and balancing the gases of our air, the atmosphere would burn itself up rather quickly. And if living things didn't turn salty nitrates into nitrogen and pump that nitrogen into the air, the seas would become too salty for life to go on in them, and the atmosphere would lose its balance.

While the Sun has been growing steadily larger and hotter since the Earth was formed, the Earth, like a warm-blooded creature, has kept a rather steady temperature. Lovelock has developed computer models showing that this "thermostat" function can be based on self-regulating changes in the reflective index of (light) cloud cover and (dark) vegetation, heating and cooling the planet's surface as necessary. In the real world, part of the complicated system involves regulating the amounts of "greenhouse gases" such as carbon dioxide and methane, which trap solar heat;

another part involves relationships between the cloud-seeding gases produced by ocean plankton and the total cloud cover of the Earth, which in turn regulates the amount of sunlight reaching the surface of Earth. Old attempts to explain how geological mechanisms might regulate the Earth's temperature are thus giving way to new explanations of how a live planet does it.

The right balance of chemicals and acid in the seas and in the soil, and even the balanced overall temperature of the Earth—all of the conditions necessary for the life of our planet—are regulated by the living planet as similar balances are maintained in our self-regulating bodies. Physiology has shown that the body knows itself as a whole—meaning, in scientific terms, that it is self-referential—which enables it to carry out its functions. It appears that this is the case for the Earth as well.

However much we learn about how the complex coordinated systems of our own bodies' functions, we will never know everything involved in building and running such systems. Our bodies work without asking any assistance of our waking consciousness, our thinking minds. Lewis Thomas has said that for all his physiological knowledge, he would rather be put behind the controls of a jumbo jet than be put in charge of running his liver for even a day. Any one of our organs is more complex by far than the most complex computer we've invented—and it knows how to run itself, repair itself, and work in harmony with all other organs.

Clearly we see Earth acting intelligently in the same way. Should we not acknowledge and incorporate this observation into biology as physiologists incorporate body intelligence into their work? The sooner we recognize and respect the Earth as an incredibly complex self-organized living being, the sooner we will gain enough humility to stop believing we know how to manage it. If we stay on our present course and cling to our present belief in our ability to control the Earth while knowing so little about it, our disastrously unintelligent interference in its affairs will not kill the planet, but it will very likely kill us off as a species.

To go back to Lovelock's Gaia Hypothesis, or Gaia theory, it is important to acknowledge a certain contradictory aspect of it which I believe has confused the issue of Earthlife. Lovelock speaks of Earth as alive but also calls Gaia a self-stabilizing mechanism made of coupled living and nonliving parts—living organisms (biota) and nonliving physical (abiotic) environments—which affect one another in ways that maintain Earth's relatively constant temperature and chemical balance within limits favorable to life. Lovelock describes this system as a cybernetic device working by means of feedback among its coupled parts to maintain Earth's stable conditions. For Lovelock, "organism" and "mechanism" are equally appropriate concepts, but in fact the two concepts contradict each other logically.

An autopoietic system is self-producing and self-maintaining. It must constantly change or renew itself in order to stay the same—your body renews most of its cells within each seven years of your life, for instance, and its molecules far more rapidly. No mechanism has ever done this, because a mechanism is neither autopoietic (self-created) nor autonomous (self-ruled) but rather is allopoietic (other-created, by an inventor) and allonomous (ruled by inventor-given laws). It cannot change itself, except as programmed by its inventor, and this is the essential difference between living systems and mechanical ones, including even the most sophisticated computers and cybernetic robots.

Thus we have to be very careful when abstracting mechanical models from observed living systems. Life cannot be part of a cybernetic device, or even part of a living being; life is the essence or process of the whole living being. If Gaia is the living Earth, then it would be as meaningless to say that life creates its own environments or conditions on Earth as it would be to say that life creates its own environments or conditions in our bodies. If we see Earth as alive, we can still say that its organisms create their environments and are created by them, in exactly the same sense as we say cells create their own environments and are created by them in our bodies. In other words, there is continual and mutually creative interaction between holons and their surrounding holarchies. But we do not divide living bodies or holarchies into "life" and "nonlife."

If we adopt my position, then, of seeing the Earth as a self-creating living entity, we also see that only limited aspects of its function—never its essential self-organization—may be usefully modeled by cybernetic systems, just as we can usefully model aspects of our own physiology (for instance, temperature regulation) as cybernetic feedback systems, but we must not mistake our models for what is modeled.

My concept of the living Earth is just that—a concept, a definition, not a hypothesis to be proven. I simply wish to replace the prevailing concept or metaphor of the Earth as a complex nonliving mechanism with the concept of Earth as a living entity by definition. Within this conceptualization we are then free to generate fruitful hypotheses about its functions or physiology, as Lovelock certainly has done despite these contradictions.

(2) Autopoiesis: the living Universe. The concept of autopoiesis thus lets us expand our view of living entities to the Earth itself, and, as I'd like to show now, to the universe at large.

Let us try seeing the Earth embedded within an autopoietic galaxy, itself embedded in an autopoietic universe. As I said earlier,

I adopted Jantsch's proposal that the universe as a whole is a vast self-creating or autopoietic living system. In a highly simplified description, my view of its process is as follows:

a) We assume the early universe expanded outward very forcibly, whether begun by the Big Bang, Bohm's explosive wave in the primal energy sea (also known as the zero-point energy field), Bentov's continual emergence of matter from a cosmic white hole (1977), or whatever else might have caused universal expansion. This out-thrusting energy/matter would have curled in on itself to naturally form whirlpools wherever unevennesses occurred, and such irregularities in the early universe have actually been found since my book was published.

b) Those whirlpools or vortices that exhibited a dynamic (non-equilibrium) balance between forceful expansion and gravitational attraction, rather than dissolving or congealing into static clouds, became the fundamental autopoietic or proto-autopoietic life forms within the whole-entities that hold stable forms over long time periods, continually re-creating themselves from new matter, eventually as spiral proto-galactic clouds. Note that whirlpools behave this way whether they exist in rivers or in outer space. Huge dynamically stable whirlpools have been discovered in our oceans, visible only from satellites. Whether short or long-lived, these vortices are the simplest forms of self-organizing entities that continually create themselves and may thus be considered proto-life entities. If one argues that whirlpools die if the river's flow cease, one also must take into account that you and I would die if our environment ceased to flow through us. That is the nature of holarchy.

c) These macrocosmic whirlpools, the protogalactic clouds, gradually evolved vast and local star systems (later also planets) from the smaller whirlpools of gases (later dust), forming within them and interacting with the microscopic whirlpools we call atoms and subatomic particles (later molecules). These huge and tiny whirlpool entities co-create increasing complexity between them in an evolutionary process that is both universal at the level of galaxies, supergalaxies, and the great clusters beyond them, and local within living planets such as Earth.

Living galaxies and stars are known to slough off parts and incorporate new material, as all whirlpools do, to maintain themselves. Some galaxies appear to swallow others up, some die and no longer give birth to stars. A star appears to be a self-organizing entity, keeping itself alive by drawing in new material and sloughing off old, with a series of life stages often ending in its explosive reproduction as a new star system including planets. The recent Hubble photographs of stars being born from great cosmic clouds remind us of budding hydras in the sea, in a beautiful new example of "as above, so below" patterning. Other photos, coming in day by day, show amazing lifelike forms we have never before seen in nebulas and other larger size scales. It is interesting that astronomers now observe that new stars are apparently born only in spiral galaxies, not in elliptic galaxies, which seem to be those that "died."

PILLARS OF CREATION IN A STAR-FORMING REGION
(Gas Pillars in M16—Eagle Nebula)

Undersea corral? Enchanted castles? Space serpents? These eerie, dark pillar-like structures are actually columns of cool interstellar hydrogen gas and dust that are also incubators for new stars. The pillars protrude from the interior wall of a dark molecular cloud like stalagmites from the floor of a cavern. they are part of the "Eagle Nebula," a nearby star-forming region 7,000 light-years away in the constellation Serpens.

The picture was taken on April 1, 1995 with the Hubble Space Telescope Wide Field and Palnetary Camrea 2. The image is constructed from three separate images taken in the light of emission from different types of atoms.

[This photograph supplied by NASA and the NSSDC. Photo No.: STScI-PRC95-44a]

Generations of stars are formed through explosion and re-creation of new whirls of "debris." In living galaxies, planets eventually form from heavier elements around new stars, and some of these self-create as extremely complex entities, such as our Earth. Our solar system was born from the scattered gases and other materials of one or more exploding super-novae more or less five billion years ago. Only one of its planets came alive and remained alive, at least from the perspective of our perceived four-dimensional universe (leaving open the possibility of life in other realms or simultaneous universes with their own dimensions).

My metaphor for the reproduction of cosmic life is that the Cosmos scatters planets as star seed, much as plants and animals here below scatter their seed. In both cases, only few seeds in this prolific venture of life actually "sprout"—those that land in the right conditions to support their continuing life. And so it makes sense that only a few of many planets flourish alive—"a few" meaning untold billions, given the scope of the universe. Lovelock has pointed to the "Goldilocks effect" on Earth: Mars too cold, Venus too hot, the Earth just right.

d) What about the tiniest size scales of the universe? One of the great mysteries of physics has been the stability of the atom. Many atoms do not decay but appear to exist virtually forever. Could decaying atoms be like dying protogalactic clouds, which become dead disks or ovals, while nondecaying atoms are more like living spiral galaxies? If so, from whence would atoms draw the matter/energy to keep them alive? Physicist Hal Puthoff proposes that they indeed draw energy continually from the zero-point energy field now known to exist by actual measurements (1990), and in personal discussion, he felt that my vortex model of the basic life form seemed to fit these atoms.

At every level of the evolving universe, then, we find autopoietic or proto-autopoietic holons nested within each other, exchanging resources with the larger holons in which they are embedded and the smaller holons embedded within them. In this sense the entire universe is a vast living entity or holon. It is especially important to note that this holarchic model gives as much importance to the downward influence of the larger holons as to the upward influence of the smaller holons. We must learn to see the world and all the Cosmos through macroscope and microscope at once!

One of the great advantages of this model is that it obviates that endlessly confusing search of Western science to determine how life can come from nonlife. We can devote our efforts to describing how a living universe evolves beyond these simple forms into such complex holons as cells, organisms, and the whole Earth in its ongoing evolution.

Within holons, there are obviously more and less highly organized, more and less complex, more and less animated parts—even dead parts, though we can count on their eventual recycling by some holon. Recall Vernadsky's continual transformation of "geological matter" into "biological matter" and back over time in a view that makes what we call "life" and "nonlife" appear to be a kind of complementarity akin to that of matter and energy. In our own developing model, let us reserve the word "organism" for multi-celled creatures, to distinguish them from cells, live planets, galaxies, etc., but let us regard what is presently seen as "inanimate matter" such as rock or water, in the same way that we regard tooth or bone or connective tissue—the "less animated" parts of a body, remembering how they are endlessly transformed at slower timescales than the rest of the crust, just as are bones and connective tissue with respect to the rest of the body.

(3) Consciousness in holarchies. Earlier, you said: "If there is consciousness anywhere in the holarchy—in the scientist-holon, for instance—then it is characteristic of the whole and we needn't be surprised to see something like consciousness appearing anywhere, especially in living organisms."

I personally believe that consciousness does indeed permeate the universe, that the universe proceeds intelligently in its evolution and must therefore be conscious. Autopoiesis as a definition of life, together with the holarchic model of living entities, makes it plausible that consciousness is inherent in every level of the universal holarchy by logical argument. Let me show how within the framework of Western science.

The autopoietic definition of a living entity (holon) as one that constantly creates and re-creates itself implies that it is an open system in nonequilibrium with a context from which it draws energy and/or matter and into which it disperses its products. Thus, living systems must be embedded within larger systems that can be seen as their environments, though they are also living systems in their own right.

This openness of material/energetic systems can be represented logically, as Walter Pankow has discussed (1976). The logical model nicely differentiates living systems from logically closed or only formally open nonliving or mechanical systems, using reasoning similar to that of Jantsch and Prigogine in their descriptions of dissipative (self-organizing and evolving) systems, but also going beyond them.

Pankow points out that all living systems are logically open, and that logical openness is equivalent to the capacity for self-reference, which Bertrand Russell and Kurt Gödel showed was not a capacity of formal (nonliving) systems such as mathematics or logic (or machinery, which is a translation of statements in such languages into matter). In formal systems or languages, statements referring to themselves can only be tautologies or paradoxes—in other words, meaningful statements about themselves cannot be made. Pankow goes on to argue that the self-reference of living systems amounts to self-representation, which also implies self-transcendence, as a system must clearly go beyond itself, to another level of logic, to refer to or represent itself. Self-reference thus implies self-perception, and self-transcendence is meta-perception, or perception of perception.

In simpler language this means that living systems can function at a logical level from which they know themselves as wholes. Self-transcendence as meta-perception permits conceptualization and categorization in holarchies of context. Thus Pankow defines the capacity of meta-perception as consciousness.

In this view, then, living or autopoietic systems are inherently self-transcendent, conscious systems, using their capacity for self-representation or "knowledge of self" in their continual creation and re-creation of self. This clearly makes them intelligent by the definition we gave above, of intelligence as the use of information to guide action. To use information you must be able to know and map the system in which you want to use it, regardless of its level of complexity. A familiar example of this inherent biological intelligence is our own bodies, which, without our conscious contemplation and intellectual assistance, repair damages to any part by drawing on tremendously varied resources, raising fevers or cooling them down, sending chemicals in highly specific dosages to highly specific locations, flushing poisons, etc., etc., ad infinitum. They could hardly achieve this without knowing themselves as wholes in all their complexity.

Perception itself can be abstracted as a more or less mechanical process if we isolate a perception from its context to trace a particular input stimulus and its effects in a particular organism, as, for example, in tracing the linear sequence of measurable light waves, electrical discharges in the eye, chemical flows in the brain, etc. In relatively simple creatures, such as bacteria, we can "attach" such sequences to the further sequences of chemical and mechanical responses or actions. But even in bacteria the situation is already too complex to describe everything that happens, because they can be observed to make conscious intelligent choices, as Lynn Margulis has pointed out (1991). So they must know themselves and their capabilities and choose among them in a variety of situations. Most impressive is the fact that all bacteria are capable of making intelligent choices of which genes to trade with other bacteria to meet new situations. This shared gene pool makes them an immense planetwide organism (S. Sonea and M. Panisset, 1983; Margulis and Sagan, 1995). In order to increase their complexity, to evolve, living systems must self-transcend, i.e., represent or know themselves as holons and explore the possibilities for change, including novelty.

As we map the living systems self-creating in the universal process, our fundamental concepts, we have agreed, will be holons in holarchies. If we adopt the model of the entire universe as a living evolutionary process generating such holarchies, this conception of self-transcendence as conscious intelligence begins to give us a very exciting "radical unity." What is so important is that this model gives us a way of talking about consciousness and intelligence not as an emergent property of evolution, as I myself described it (1989, 1996), and as Roger Sperry and others see it, nor as Bergson's separate life force, Alfred North Whitehead's creative principle, nor Sir Arthur Eddington's and Wald's ideas of consciousness permeating the universe yet being separate from matter. Rather, it depicts conscious intelligence as a fundamental feature of every holon of a living universe from its inception.

Erich Jantsch defines "mind" as the self-organizing dynamics of a system; Pankow shows us the logical details; Gregory Bateson said that epistemology itself is the integrated meta-science of mind, which has evolution, thought, adaptation, embryology, and genetics as its subject matter.

Note that this view of an essentially alive and intelligent universe gets us around the sticky problems remaining in Fred Hoyle's notion of panspermia with some unknown but particular life source within an otherwise nonliving universe; also Arne Wyller's "localized Hoyle proposition" that we identify such a source as our own intelligent planet (1996). Wyller asks, in comparing the efforts of artificial intelligence designers with natural evolution: "If our best minds cannot beget intelligence, how can chance do it?" (p. 215) He further asks whether it is reasonable to think intelligence can spring from nonintelligence. The view I have described shows intelligence, as well as the life it implies, to be present from the beginning.

HARMAN:

Whew! I am enthused about your conclusions. I must admit, however, that simple arguments are even more persuasive to me than the very complex ones.

The revolutionary issue that we are dealing with is the ontological stance earlier identified, namely that reality is considered to be holons within holarchy rather than fundamental particles. If that is accepted (and why not, since it is not in conflict with any of our experience), then much else follows easily. If there is consciousness anywhere in the holarchy, then it is characteristic of the whole and we needn't be surprised to see something like consciousness appearing anywhere. Autopoiesis seems clearly to be related to that "something like consciousness." We can come to understand it because it is of us. The much-disparaged realm of mystical experience is in essence one aspect of how humans have contacted this knowing; the direct knowing of indigenous peoples is another; the realm of aesthetic creativity is another.

Taking this approach, consciousness is inherent in every level of the universal holarchy and does not need logical justification. The record of human experience is found there, and we need no longer cast doubt on such reports because they are not in accord with the self-limiting Western scientific worldview.

In other words, rather than search for ways within the Western scientific paradigm to support the contention of consciousness permeating the universe, it seems a more straightforward argument to observe that Western science is based on metaphysical assumptions that we have reason to question, and hence we should explore alternate ontological and epistemological assumptions.

SAHTOURIS:

That very simple view has an appealing elegance, and it is of course the view that all so-called "primitive" cultures, past and present, have held and still hold because it is what they directly experience. They know through their own perceptions that all nature is alive, intelligent, and conscious. We of the Western scientific/technological society are expected to make good logical arguments for such a position because we are taught to see all nature other than ourselves as a collection of "things" to be studied, manipulated, and used to our ends. We define their attributes through our own set of tools, including mathematics, logic, and mechanical instruments. What matters at this point is that we come to the same conclusions from both perspectives, because that has the greatest power of persuasion. The fact that this is happening is very exciting to me. It not only increases our respect for these other cultures, as we see that they are far more sophisticated than we had perceived them to be, but gives us a newer integrated worldview and a science that we could not have achieved solely through our own limited perspective.

GLOSSARY

Allonomy: (this one coined by Jantsch) law or rule from outside, the nature of invented machinery. (In the Cartesian worldview, nature was ruled by its inventor, God as the Grand Engineer.)

Allopoiesis: literally other-creation; a term useful in distinguishing machinery, which is always created by an inventor who determines its rules of operation from the outside (see allonomy).

Autonomy: law or rule *(nomos)* of the self *(autos);* evolved from within.

Autopoiesis: literally, self-creation; our first central definition of a living entity as anything that constantly creates itself. Note that this implies a context from which it can draw matter and/or energy and into which it can release its products, including wastes.

Ecology: organization *(logos)* of the household.

Economy: law or rule *(nomos)* of the household *(oikos)* with "household" interpreted as a living entity such as a cell, body, ecosystem, or society. (Note that these two definitions show the inherent inseparability of economics and ecology, a clue to the problems created in human attempts to do so.)

Evolution: the improvisational dance of nature over time, in which workable steps are kept while new ones are evolved toward the health and creative expansion of complexity in the whole.

Holarchy: the embeddedness of living entities within each other (e.g., cell, organ. body, family, community, ecosystem, bioregion, planet, star system, galaxy, etc.).

Holon: a living entity or system.

Holonomy: law of the whole that comes from the self's context, the holons in which it is embedded.

Mutual Consistency: the dynamic (not static) harmony resulting from the interplay of holons' self-interest at every level of their holarchy (e.g., the self-interest of an organ and of its body, or of person and partnership, are worked out in mutual consistency)—perhaps the fundamental operating principle of the universe.

Part/Whole Tension: the fundamental source of creativity at all levels of holarchy.

CHAPTER
FOUR

Towards A Holistic Biology

Scientific distinctions between mechanics and organics have been blurred because we ignored the fact that mechanisms are, by definition, the purposive constructions of their inventors, and therefore cannot exist as natural entities evolving in purposeless (non-teleological) nature. Our whole scientific concept of nature as mechanism was derived from a Cartesian scheme that was logically complete because it included God as inventor. But to maintain that nature is mechanism after repudiating God and purpose constitutes a severe logical flaw at the heart of Western science.

—Elisabet Sahtouris (1996)

HARMAN:

What we are discussing suggests such a radical transformation of biology that I wonder if we shouldn't pay some more attention to biologists who recognize the troubling puzzles in biological theory but propose less drastic ways of dealing with them. I'm thinking, for example, of the structuralism of Brian Goodwin and Mae-Wan Ho, and the "dialectical evolution" concept of Lewontin and Levins.

Dialectical evolution. As Levins and Lewontin say (1985), in the Cartesian world, "phenomena are the consequences of the coming together of individual atomistic bits, each with its own intrinsic properties, determining the behavior of the system as a whole. Lines of causality run from part to whole, from atom to molecule, from molecule to organism, from organism to collectivity. As in society, so in all of nature, the part is ontologically prior to the whole." This view dominates Western science. But in fact, parts and wholes evolve in consequence of their relationship, and the relationship itself evolves. Parts and wholes have a dialectical relationship: one thing cannot exist without the other, and one acquires its properties from its relationship to the other.

Two pertinent examples of this dialectical nature of evolution are co-evolution of the organism and its environment, and the co-evolution of science and Western society. "Science is a social process that both causes and is caused by social organization." The bourgeois revolution in Europe, and the replacement of hereditary holders of power by those whose power derived from their entrepreneurial activities, was accompanied by growth in the ideology of change as an essential feature of natural systems. Thus, they say: "An evolutionary world view is really congenial only in a revolutionizing society."

"Evolution is neither a fact nor a theory, but a mode of organizing knowledge about the world." Levins and Lewontin assert that the concept of order in the evolutionary process is an ideological one. Entropy gives directionality to inorganic systems; there is no such mathematically defined measure to give directionality to evolution, although anti-entropic directionality (toward order, rather than disorder) seems characteristic of both life and evolution. For nineteenth century evolutionists, evolution meant *progress*. Stability and homeostasis were emphasized in late 19th century evolution. "Theories of evolution are now literally preoccupied with stability and dynamic equilibrium." The concept of adaptation in evolution is that nature sets "problems" and evolution consists in "solving" these problems—as the eye solved the problem of sight, wings of flying, etc.

By contrast, their principles of a dialectical view are: (pp. 273–274)

1. A whole is a relation of heterogeneous parts that have no prior independent existence *as parts*.

2. The properties of parts have no prior alienated existence but are acquired by being parts of a particular whole.

3. The interpenetration of parts and wholes is a consequence of the interchangeability of subject and object, of cause and effect. Example: Organisms are both the subjects and objects of evolution—they both make and are made by the environment and are thus actors in their own evolutionary history. (For example, as we have discussed earlier, the reducing atmosphere that existed before the beginning of life has been converted, by living organisms themselves, to one that is rich in reactive oxygen.)

4. Because of (3), change is a characteristic of all systems and of all aspects of all systems.

The assertion that all objects are internally heterogeneous leads us in two directions. The first is the claim that "there is no basement"—no "fundamental particles." This argues for the legitimacy of investigating each level of organization without having to search for fundamental units. A second consequence is it directs us toward the explanation of change in terms of the opposing processes united within that object. The parts or processes of a whole confront each other as opposites, conditional on the whole. "What characterizes the dialectical world, in all its aspects, is that it is constantly in motion. Constants become variable, causes become effects, and systems develop, destroying the conditions that gave rise to them." It is not change alone that requires explanation, but persistence and equilibrium.

Radical unity. I find it very exciting to feel that we are now coming to a more holistic view—to what British biologist Brian Goodwin calls "radical unity." In this view, becoming and knowing are inseparably joined. Witness the alchemist's experience of gnosis as he (subject, observer) and the material in the crucible (object, observed) undergo mutual, appropriate transformation in a relevant praxis without which there can be no licit knowledge. This, to the alchemist, describes our sacred dance with reality. Goethe had a similar view of science, and Henri Bortoft has recently described this kind of biology in modern terms in his book *The Wholeness of Science* (1996).

As Goodwin has noted, his "radical unity" is very similar to the sixteenth century worldview portraying the union of becoming and knowing, of ontology and epistemology, of the knower and the known, of consciousness and matter. The dominant worldview of the 16th century assumed a deep unity between nature and gnosis (knowledge, hidden but accessible to imaginative thought and feeling). What emerged in the early seventeenth century was a science based on a profound division between mind and the nature it contemplates. An "ontological gulf" came to exist between consciousness and its object. The "real" became an object which stands over against the thinking mind, appearing to it but not in it.

This dramatic change in perception emerged from a fierce struggle around the beginning of the seventeenth century between two groups championing radically different conceptions of development and reform in Europe. This struggle for legitimacy and power, between Cartesian dualism and Renaissance nature philosophy, was a fateful event in Western history. The holistic medieval worldview was shattered. As the contemporary poet John Donne mourned, "'Tis all in paeces, all cohaerence gone."

SAHTOURIS:

Both the dialectics of Levins and Lewontin and the radical unity of Goodwin make sense as aspects of revisioning biology. And it should already be obvious that I agree that the theorizing and research of the whole British group around Goodwin and Ho is a most promising direction in biology today. Goodwin does, as you say, take a very radical step by going back to a "prescientific" European worldview to propose a participatory biology that would again close the gap—the "ontological gulf"—between scientist and phenomenon, consciousness and its object, being/ontology and knowing/epistemology. However, I believe that this step backwards is the only way forwards. We've gone as far as we can with these dualisms, and perhaps we can count that journey into reductionist/materialist "objectivity" as a productive historical digression. But our separation from outer nature, and from our own inner natures, in this technological world spawned by science, now endangers our very existence to the point where we are literally forced to reconsider the baby we threw out with the bathwater—the living, participatory-whether-we-like-it-or-not universe.

Levins and Lewontin's observations on the dialectic of evolution and their acknowledgement that "Evolution is neither a fact nor a theory, but a mode of organizing knowledge about the world" are both relevant in the transition to a new biology. Evolution biologists can be seen as making maps of the temporal processes of Earth's living systems and, as I've often said, different maps to the same territory are useful for different purposes. This "dialectic evolution" map is similar to Lovelock's cybernetic model—the biota modifying the abiotic environment and vice versa. Levins's and Lewontin's "interpenetration of parts" model, like Lovelock's cybernetic coupling of biota with abiotic environment, is surely preferable to the one-way Darwinist concept of the environment shaping species.

Other biologists will map this interactive process in their own ways. I like to call it simply the "co-evolution" of the whole—all species, all geobiological processes. The question is whether we will continue to see this as updated neo-Darwinism or whether it will move us into a truly new ways of seeing nature as a single intelligent whole. It is of some interest in this connection that Horgan (1996) cited both Stephen Gould and Stuart Kauffman as being ambivalent about whether their work constituted an elaboration of Darwin's work or an alternative to it.

Let's look at evolution in our terms—as a holarchy of holons in continuous striving for dynamic mutual consistency. Ancient bacteria engage in endless trading of genes that leads to new physical features and lifestyles as they continually recycle and re-create atmosphere, soil, and seas. Larger species (plants and animals) evolve by working out predator/prey relationships that keep each other healthy, as well as cooperative ways of sheltering each other and serving each other's reproductive needs. Ecosystems develop via total recycling; the products we call "wastes" ensure species survival by being useful to others. Species with good, resilient designs, such as squids, salamanders, and sharks, survive like bicycles in a jet age while others reach beyond themselves in bursts of new creativity. Species that do not reach mutual consistency with others may dominate and destroy for a while, but sooner or later they are starved out by their own gluttony.

As the Greek philosopher Anaximander said a few thousand years ago (by my own translation into English): "Everything that forms in nature incurs a debt, which it must repay by dissolving so that other things may form." What a marvelous description of evolution-through-recycling in a single sentence! Even galaxies, by clashing and swallowing each other, combining and separating, giving birth to stars and planets in the course of their lives, now can be seen as macroscopic participants in this improvisational dance of mutual creation, or co-evolution.

What you and I are trying to do, because we find research results as well as everyday observation of nature in conflict with current scientific maps, is to explore the development of this alternative mapping system. Whether it proves useful or not will of course depend on who decides to test it and how it holds up under such test.

In our model or map of living systems, we choose to see them as holons in holarchy engaged in these endless creative dynamics toward mutual consistency. This has been for me a coherent and meaningful way of organizing human observations of Cosmos, Earth, and its creatures. Yet, for a long time I deliberately remained within Western scientific tradition by rearranging the accepted data of science into this new model without bringing in consciousness except as a gradually emerging evolutionary phenomenon. That gave me a model that shows all life as a single interconnected, interdependent process. But it remained just that—a coherent new way to see—not a new way to explain what we see.

To attribute intelligence to the entire holarchic universe now seems an obvious next step—not dualistically, but simply as an observed property of all autopoietic living systems or holons. One couldn't do that with the mechanically modeled nonliving universe. The intelligence perceived by the "fathers of science" had to be located in a *deus ex machina,* or God outside the great mechanism—God its inventor, or, as Descartes called him, the "Grand Engineer."

When this Grand Engineer was repudiated in favor of random accidental process, meaning and purpose was stricken from the scientific worldview. Western science became content with detailed description—it had only to say "how" things were, as it denied there was any "why." The nonliving entropic universe, still framed in mechanical terms as a great machine running down to heat death, had no more meaning than a Dada machine. Less, in fact, since Dada machines had inventors with the paradoxical purpose of creating something nonpurposive or meaningless. The entropic universe, all we call "nature," simply *was,* and thus all science could do was to describe its parts, and to abstract out and copy those aspects of it useful in developing human technology.

Those philosophers of science, such as Bergson and Whitehead, who could not deny the exuberant and intelligent life of cosmos and planet, were forced to inject an *élan vital* or a divine creative principle into nature. Modern science simply ignored these philosophers, or declared them vitalists or dualists and cited them as examples of the anthropomorphic heresy.

But intelligence, consciousness, and intention still lurk insistently, as you have repeatedly pointed out, behind all the data of micro- and macrobiology. There comes a time when denying it appears as ridiculous as the scientist denying consciousness and intelligence in his own children, as B. F. Skinner did. It is like a game scientists have been taught to play, the rules of which they teach each new set of graduate students they initiate. One simply cannot break these rules without threat of being kicked out of the game. Now, however, we have come to a time of tremendous transition—a time of "saltation," in evolutionists' terms. The rules have to be broken because they no longer serve us.

Our holarchic model is perfectly amenable to seeing consciousness as intrinsic to nature, since it is a participatory model in which all is alive. Living entities, or holons, exhibit self-reference, perception, intelligence; in fact they could not exist without perception of self and surroundings, or without intelligent use of the information perception provides. An autopoietic entity is necessarily and intrinsically conscious.

How does this fit Goodwin's radical unity? In a holarchy, doesn't the "knowing" become manifest as the "known"?—isn't consciousness the source of morphology, of materializing pattern? Aren't ontology and epistemology a single process emphasizing now the one aspect, then the other? This is the concept of dynamic wholes.

HARMAN:

Indeed, Cartesian dualism in science took apart the whole in its effort to explain everything "objectively," with intelligent observers separated from the natural world of "objects" under study. Thus we have come to see the organism itself as a dualistic mechanism, one part in control, like computer software or the coding in the DNA; the other part passive hardware, like the computer itself or the body.

There is a strong tendency in modern biology to assume that organisms have an "intelligent" controlling aspect and a passive, quasi-inanimate, controlled aspect. The controlling part embodies the essential principles of the biological state: the capacity to keep the parts working in relation to one another, to reproduce, to evolve, and to adapt. The controlled part is explainable strictly in terms of physics and chemistry. The organism is seen as a dualistic mechanism, comparable to a computer with its "intelligent" software and its passive hardware that responds to instructions coded in the program.

This is precisely the metaphor used in molecular biology to explain the relationship between the DNA, with its hereditary instructions, which are intelligently translated into proteins via the genetic code, and the organism, which is made out of these proteins and their products. DNA, RNA, and protein types of macromolecules are distinctive to organisms; their relationships, and their changes during embryonic development and evolution are successfully described by these metaphors. The techniques of molecular analysis include some that exploit basic macromolecular processes distinctive to organisms, such as immunological (antigen/antibody) reactions, DNA replication (gene cloning), and DNA/RNA interactions (hybridization).

There is absolutely no doubt about the power of these analyses and techniques. What remains elusive is precisely what these techniques and the associated conceptual structure are unable to address directly, namely, the nature of the integrated spatial and temporal order that gives organisms their distinctive attributes, particularly their morphology and their behavior.

Contemporary biological explanations tend to describe only necessary conditions. They tend to ignore any reference to such laws of organization of biological systems as might be comparable to the basic laws of physics, and so deal only with necessary, not sufficient, conditions. For example, one often finds statements to the effect that mutant genes "cause" particular types of change of form or morphology in organisms. An example is a homoeotic mutant called *antennapaedia* in the fruit fly *Drosophila,* in which legs appear during the embryonic development of the fly where antennae would ordinarily arise. However, this is cause in neither a specific nor a sufficient sense. It is not specific, because the effect of the mutant gene can be produced in normal (nonmutant) flies by a nonspecific stimulus, such as a transient change in the temperature to which the embryo is exposed at a particular time in its development; and it is not sufficient, because knowledge of the presence of the mutant gene is not enough to explain why the morphology changes as it does.

The particular metaphor that biologists have used to describe biological reproduction, namely that the reproductive process is directed by a program written in the nucleic acid message, has two dubious consequences. First, it implies that genes and the "information" they contain constitute specific, sufficient explanations of reproduction, so that biologists seek solutions to the problem of reproduction in terms of hierarchies of regulator or control genes and the "sub-routines" they control. This splits the organism into genotype (instructions) and phenotype (the form produced by the instructions); as we have remarked earlier, that split led to the dogma that there can be no transfer of information from organism to genome. (This split, we recall, was introduced by Weismann in 1894.)

The second consequence is that this metaphor also makes biology seem basically different from physics, in that it is primarily a science of information processing rather than one characterized by particular types of organization of matter and distinctive fields of force.

In the mechanistic view of biology, equilibrium is expected; change is to be explained. By contrast, in a process view, change is expected and stability to be explained. In the holistic view (immanent causation), "things are doing what comes naturally." For example, one can't find *external* causes in embryogenesis; the organism is causal to itself, it has *aliveness*.

An organism is always *becoming:* in order to be true to its own nature, the organism has to change and transform constantly. Goodwin insists that to understand life we have to replace the ahistorical, reductionist paradigm of modern biology with a generative paradigm of process. The quintessential characteristic of organisms is dynamics. We shall never understand life through looking at the fixed shape; the single leaf or flower is only a milestone in the plant's development, a visible product left by life flowing on. We have to start with a concept of the whole organism and its life cycle. *This is the fundamental entity in biology;* it generates parts that conform to its intrinsic order.

SAHTOURIS:

You've illustrated beautifully the problem with the mechanistic view: It freeze-frames what is inexplicable apart from its dynamics and then further confuses the issue by dividing this static form into parts, including this current biological model of the organism as a DNA-software programmed computer. Scientists are notorious for believing nature to emulate our latest techno-logical inventions. Recall Freud's plumbing model of the brain, where things got jammed up and valves had to be opened to release them, etc. This was followed by the telephone switch-board model, then the computer, the holographic camera, and now the parallel processor. We get increasingly more sophisti-cated in copying nature, reducing some of nature's functions to our abstracted mechanical assemblies, but however much we then proclaim nature to have copied our inventions, it remains the other way 'round. The brain is not any kind of computer, nor is a single cell "programmed" by its DNA, because neither of them are mechanisms assembled from parts by some external inventor.

Confusions between mechanism and organism also bring up for me the problem of measurement and modeling in the study of living systems. You've talked about consensual validation and other alternatives for assessment of research results, which I believe we will necessarily see much more of, but to a consider-able extent scientists will continue to reduce data to equations and use computer modeling in such mathematical languages as nonlinear equations, dynamics and cybernetics. We need to be aware of how much we can actually learn from such models and what their limitations are. Brian Goodwin, for example, has made fascinating progress with his computer models of morpho-genesis, but there are some interesting questions about the limits of such models. While they can abstract geometric patterns or cybernetic interactions, they break down very quickly when any kind of natural complexity is introduced, and they have not been successful with modeling the evolutionary process for this rea-son. This brings us to the broad question of the extent to which mechanics can model organics.

Let's review the fundamental differences between living and mechanical (formal) systems: We earlier discussed them as logically open and closed systems (Chapter Three), noting that formally closed systems cannot represent themselves, as Russell and Gödel showed. Scientific distinctions between mechanics and organics have been blurred because we ignored the fact that *mechanisms are, by definition, the purposive constructions of their inventors,* and therefore cannot exist as natural entities evolving in purposeless (nonteleological) nature. Our whole scientific concept of nature as mechanism was derived from a Cartesian scheme that was logically complete *because* it included God as inventor. But to maintain that nature is mechanism after repudiating God and purpose constitutes a severe logical flaw at the heart of Western science. This is an extremely important matter that one simply does not see discussed, as it seems to me it must be.

Now that we can define and see life as autopoietic, we can also see more clearly that machinery, by contrast, is what I have called *allopoietic* (other-created), with a different inherent logic. Mechanisms, and the mathematical/logical languages they are based on, are not self-referential and cannot understand themselves as wholes; thus they are not and cannot, by definition, be holons. They exist within the holon of humanity as invented extensions of humans, transformations of matter from the human holon's environment. To see the difference between mechanics and organics more intuitively, recall that living holons maintain their continuity through a continuous renewal or exchange of materials in all their parts, while mechanisms depend on their creators or maintenance people for positive change. If you walk away from a mechanism, you hope, in fact, that it does not change, since machinery is not likely to improve itself in your absence; on the other hand, if you leave a friend or other living entity, you hope it keeps changing in your absence, lest it die.

Since ancient times, when formal languages were first constructed, they have been required not to change over time, culture, or whatever natural language was used to teach them. Thus they had to be built of elements and rules for combining them that would eliminate the changing, living aspects of natural languages. In other words, any capacity of natural language for self-reference and self-transcendence was removed in constructing artificial languages, whether this was a conscious process on the part of their inventors or not. The boundaries of formal languages separate them, while the boundaries of living systems join them. Note also the one-way embeddedness of formal language in natural language: Math or logic must be taught in a natural language; one quite obviously cannot teach a natural language with math or logic.

These differences make it clearer why computers, despite the fond hopes of artificial intelligence designers, do not translate living languages well by intelligent literary standards; for example, they are unable to write or even translate poetry. Paul Watzlawik, and later Walter Pankow, claimed that living systems can only be represented adequately in living (natural) languages, which are themselves self-transcendent. In a scientific experiment, we create a setup for the measurement of relationships among certain (very few) separate variables abstracted (isolated) from their usual context and hope this will explain a phenomenon outside the experiment. Measurement is an indirect observation that necessarily excludes qualitative factors. We literally transform the natural phenomenon under study into a mathematical system to arrive at mathematical functions: formal relations or "laws" between independent and dependent variables, often in a logical "if . . . then" causality. In this way science transforms the world into its own image and becomes a self-fulfilling prophecy. As Pankow points out: "The explanation is no longer geared to the phenomenon, but the phenomenon is adapted to an already existing explanation."

With more holistic or nonlinear mathematical systems, such as dynamics and computers, the models are capable of far greater complexity than laboratory experiments, but actual measurement of each variable becomes impossible, so the models themselves become the experiment (as simulations), and decisions about their validity are based on "fit" with direct observation of the real world, with limited measurement on a few variables to confirm this fit. That is, nature can't be made to fit the model, so we fiddle with the model until it best reflects reality. The essence and qualitative aspects of life will remain elusive, and this argues strongly for seeking new means to evaluate qualitative living-system phenomena.

HARMAN:

I think you are quite right about this problem of measuring and modeling. It points in the same direction as the bias of Western science that I noted earlier—the assumption that a nomothetic science, one characterized by inviolable "scientific laws" relating to quantifiable variables, can in the end adequately give us the understanding we seek.

In evolution there is a generative order that is necessary to understand; random mutation and natural selection are there, but these are not sufficient to explain form. The deep intelligibility of nature is to be seen in the kind of order that produces the chemical elements as stable states involving different configurations of the fundamental particles. Similarly in evolution, species are the "stable states" reflecting deep rules of order and organization.

There is a continual attempt in biology to reduce form to molecular composition, but it doesn't work. As an analogy, one might attempt to deduce the form of spiral flow of water down a drain from its molecular composition, but it won't work; all liquids exhibit similar form, which can be described in terms of hydrodynamic principles and field equations.

Separating intelligence or consciousness from "biological mechanisms" in a reductionist science has led to a situation in which consciousness has been pushed almost outside the realm of legitimate scientific study. One of the most important contemporary developments, it seems to me, is the growing recognition that consciousness has been so neglected in Western science, and must now be brought in.

A particular form of the holistic view is found in the structuralism of Goodwin and Ho. According to them, the explanation of development (ontogeny) is not to be located solely in the genes. Instead of a one-way transfer of genetic information, there is a whole network of feedback interrelationships between organism and environment. Heredity does not reside solely in the DNA; it does not inhere in any particular material substance passed on from one generation to the next, but rather it is a property arising out of the same nexus of interrelationships that catenate within and between generations. The organism is an integrated whole, genotype with phenotype, body with germline. These wholes are themselves in continuity with past and future generations through the nexus of physiological, ecological, and sociocultural relationships.

This structuralist view assumes that biological phenomena are intelligible in terms of expressing a particular type of order. That is, phyllotactic (leaf) patterns in plants, tetrapod limbs, and eyes are interpreted as natural consequences of the dynamic order of the living state. The basic principle of structuralism is that *function precedes structure*. Research of Goodwin, Ho, et al. suggests that biological forms are a result of what comes naturally from the dynamic properties at work in developing organisms; they amount to solutions of morphogenetic field equations with moving boundary conditions. Morphogenesis is not the result of a "genetic program," but rather, the consequence of the dialectical relationship of dynamics generating geometry and geometry modifying dynamics. The genes define the range of parameters within which these developments take place, but the organizing principles of the process are embodied in the spatiotemporal properties and behavior of the cytoplasm/cell wall morphogenetic field. It is not sufficient to know the molecular composition of these structures to explain form; it is necessary to understand the organizing principles. This approach also helps understand how form can be influenced by calcium concentration in surrounding sea water, electrical properties of the surrounding medium, etc.

Genes define necessary, but not sufficient, conditions to explain form. The structuralist belief is that there are generative principles operating in organisms that give them an intelligible unity encompassing their obvious diversity. Furthermore, these principles can be discovered by appropriate research programs.

Goodwin gives an example from the world of plants. There are about 250,000 different species of high plants—the ones that are familiar to us, with roots and stems, green leaves, and flowers. Despite the profusion of different leaf shapes in these plants, there are basically only three types of leaf and flower pattern: whorls, alternating, and spiral. This leads to the proposition that there are "only three basic morphogenetic attractors for the dynamic organization of the meristem [the growing tip of the plant] as a growing system with a moving boundary. . . . The three basic phyllotactic patterns [arrangements of leaves on a stem] are the stable morphogenetic solutions of [the] mechanical strain fields. . . . Just as no one uses quantum mechanics to understand the strength of a particular bridge construction or the buckling of a beam, so no one should expect to explain phyllotaxis in terms of hormones or gene products. It is the wrong level of analysis. This is another of the precepts of structuralism: model at the appropriate level, and make no assumptions about causal reductionism, or preferred levels of explanation." (Goodwin, 1994a)

The eye is an often cited example of a problem with the neo-Darwinist premises. How could random variations ever fortuitously come together to produce the first elementary eye, from which natural selection could produce the highly refined visual systems found throughout the vertebrates and in such invertebrates as gastropods, cephalopods, crustacea, and insects? What is even more amazing is that eyes of various designs appear to have evolved independently many times—perhaps as many as forty times. But Goodwin claims (1994a, p. 162) that from a structuralist point of view, "eyes are not improbable at all. The basic processes of animal metamorphosis lead in a perfectly natural way to the fundamental structure of the eye." He shows how a primitive but functional system for recording visual images could have arisen independently in many taxa as a natural consequence of the dynamic principles that operate in animal embryos involving the folding, spreading, and interaction of sheets of cells. Such a development would be a "first necessary step in the evolution of more sophisticated visual systems, which arise by extensions and refinements of basic morphogenetic movements.

The processes involved are robust, high-probability spatial transformations of developing tissues, not highly improbable states
that depend on a precise specification of parameter values (a specific genetic program). . . . Eyes have arisen independently many
times in evolution because they are natural, robust results of
morphogenetic processes." (pp. 167–168)

Here we have an example of three potential levels of explanation, all complementary, no one contradicting another: molecular (genetic information), structural (dynamic principles), and
creative emergence (eyes evolved because the potentiality of
sight was recognized by underlying intelligence). The first of
these is well accepted and, by some scientists, apparently considered to be sufficient. The second is less universally appreciated,
but would generally be considered to be a legitimate area for
exploration. The third is typically considered to be vitalistic
nonsense, long discarded by the mainstream scientific community. Yet as I think we shall see (Chapter Five), it may yet have a
place. The more complete explanation, we are suggesting, is
something like this:

Genes ↔ Morphological field ↔ Creative emergence

As Brian Goodwin has insisted, there is no adequate meta-
physics of process on which to build a scientific theory of organisms as agents manifesting immanent causation. Organisms are
not billiard balls knocked about by random mutation and shaped
by natural selection. They are self-organizing processes engaged
in characteristic patterns of transformation, both ontogenetic
(developmental) and phylogenetic (evolutionary). They are the
causes and effects of themselves, undergoing constant change in
order to be themselves.

One of the distinguishing characteristics of a holistic biology is in the metaphors used. The reductionistic neo-Darwinism literature is replete with such metaphors as "information" in the DNA, genetic "program," "competitive" interactions among species, "survival of the fittest," "selfish genes," and survival "strategies." In the neo-Darwinist view of evolution, species either work, and hence survive, or they don't; they have no intrinsic value or holistic qualities, and the metaphors reveal this. In a holistic biology, by contrast, we find such metaphors as continuum, cooperation, altruism, creativity, agency, and intentionality. Apparent purposefulness (ostensible teleological influence) is not necessarily something epiphenomenal, to be "explained" in terms of chemistry and physics, but an observable emergent quality to be included in theories at the higher hierarchical (holarchical) levels.

All of this seems to lead to a hierarchical structure of science: physical, biological, and human. The biological sciences are limited, in general, by constraints imposed by the physical sciences, but some concepts in the biological sciences cannot be reduced to the physical sciences. Similarly, the human sciences are limited, in general, by constraints imposed by the physical and biological sciences, but some concepts in the human sciences cannot be reduced to the physical and biological sciences.

Such approaches as those of Levins & Lewontin and Goodwin & Ho deserve attention, and yet it seems they don't quite handle everything. They don't go deep enough in their analysis of the problem of fitting biological observations into a Western science framework.

I think that such biological "puzzles" will look quite different when we combine the holarchic ontology with the "radical empiricism" of William James, a combination which suggests a radically new cosmology.

SAHTOURIS:

You are so right that biology has been unduly constrained by the physical sciences, including the forced effort to build up biological entities from physicochemical components, as though they were assembled mechanisms. One of my favorite metaphors is that nature's dynamic process is far less like engineering than like mothering. She doesn't just assemble things but takes a little here and puts it there, endlessly and creatively juggling needs and resources to make everything come out well for everyone in the whole family and community. This is just another version of evolution as the improvisational participatory dance of every being comprising nature. It seems very compatible with Brian Goodwin's notion of natural beings as "causes and effects of themselves, undergoing constant change in order to be themselves."

The dancer is always there, repeating good steps, weaving them into evolving new transformations of her body in response to her audience and setting. "Mother" and "dancer" are, of course, anthropomorphic metaphors, though I would be as happy visualizing the dancer as a dolphin or a river, since the underlying metaphor is the dance (function or process) itself. We could say the dance dances the dancer (structure), as you and I and Goodwin, not to mention some quantum physicists, all agree. That dance is no more and no less than consciousness expressing itself, as I described earlier. Some call it the dance of Shiva.

Our holarchic, living-system paradigm makes it possible to see that all cells and organisms, from micro-organisms to the Earth itself, co-evolve, with intelligent communication at all levels. And, as you say, each level of the dance has its own characteristics. In terms of the dancer metaphor, we can study the minute movements of each body part, their effects and causes in the whole integral body-in-motion, the perceptions and communications among parts, the psychic/mental field and feelings of the dancer, the evolving patterns of the dance itself, and so on. It doesn't matter whether we go from small to large or large to small; it matters that we take all levels into account as equally important, and that the holon, or whole, is our reference point.

Let me try to put together some of the fundamental features or organizing principles we can observe in holons, be they cells, bodies, families, communities, ecosystems, or the Earth itself. The examples given in this list are intended to help identify problems in human systems resulting from failure to act in accordance with the systemic functions of healthy natural systems.

Self-creation (autopoiesis): All living systems are continually building and renewing themselves. In the body, for example, the molecules and cells of different parts renew themselves at different rates (stomach lining cells and the molecules of brain cells renew themselves in hours or days; other kinds of cells are replaced more slowly, but about every seven years the body is made entirely of new materials). Human families, as larger living systems, show renewal not only within their individual members but in the the way their material extensions (possessions such as home, clothing, car, etc.) are exchanged over time. The members, and so the systems as wholes, also create and exchange knowledge, thoughts, moods, and actions. Holons are in endless flow dynamics.

Embeddedness (holarchy): Self-creation requires environmental sources of matter and energy. Even protolife forms such as atoms apparently continually self-create by drawing from zero-point energy and releasing spent energy into this background; river whirlpools and thunderhead clouds similarly draw from their environments to maintain their forms, as do all cells, fungi, plants, and animals. Complex living systems are embedded in larger living systems: cells within bodies, bodies within families, organizations, communities, bioregions, human world, planet, cosmos. Notice that any individual is very deeply embedded and can only have an illusion of separateness. Autonomy is continually modified by holonomy (see Glossary, p. 129–30).

Transformational flow: Living systems take in matter, energy and information; they use and change them, then put out transformed matter, energy and information. An animal as an individual living system, for example, takes in food, water, and air, as well as sound, light, taste, smells, and other perceptual energies and information. A human also takes in information from other people, radio and TV, printed matter, computer programs, etc. All this input is transformed. Food may be cooked, eaten, digested, assimilated, transformed into muscle or brain power, and excreted; solar energy is transformed into ATP; news may be transformed into emotion; anger into love.

Complexity: Living systems have multiple parts and diverse aspects functioning in various roles. In good health this complexity functions harmoniously. The complex relations of functions and structures in even the simplest cell have yet to be understood in their entirety. In healthy human organizations, for example, we need visionaries, entrepreneurs, implementers, administrators, monetary managers, integrators, etc.

Consciousness: A living entity has the capacity for self-reference, which means perception and metaperception (the ability to recognize itself as a whole), which is equivalent to awareness or consciousness. Species within ecosystems appear to have some form of consciousness of their holarchies.

Communications: The parts of a healthy living system know each other and share and exchange information. Every cell in a multicelled creature contains information about every other cell, through the DNA it contains. Cells and organs constantly exchange messages and materials delivered to and collected from receptor and emitter sites. The species of an ecosystem have myriad ways to communicate each other's presence and states of being.

Equitable Economics (See Sahtouris, 1997): The holon knows what products are needed and to which parts they must be distributed. Production and distribution of products (such as ATP or blood or plant hormones) within individuals is shared equitably and never hoarded or concentrated when need exists anywhere in the entity. Such economics can also be seen in ecosystems where various species provide food, products, and services for each other—trees shelter birds and insects, bees pollinate flowers, mammals package seeds in fertilizer and distribute them, fungi and plants exchange materials, saprotrophs, whether microbes or vultures, recycle, birds warn of predators, etc. A healthy holon produces only quality output: all the matter, energy and information it puts out is useful to other holons. No species other than humans create waste materials that cannot be used as food by some other species. All ecosystems recycle virtually all of their materials, while it is a new concept to humans.

Mutual consistency politics: The parts of a healthy holon contribute to each other's welfare in a balance of interests, a balance of conservatism and change. Every cell looks out for itself as well as for its tissue, organ, and entity. It does not choose between "right" and "left," between conserving what works and changing what doesn't: it simply does both as appropriate to ensure its own health. Nervous systems have evolved as governments based on service to the whole: collecting and distributing information equitably, ensuring healthy function by coordinating assistance to parts in need or the healing of the whole. Misdemeanor and punishment are unknown, equal rights and responsibilities prevail.

Learning: Holarchies are always engaged in using feedback to modify function and structure. Genes are stored against future need like books in libraries or computer information banks. *Memes* are social versions of genes, i.e., ideas shared within generations and passed on to future generations. Learning tells a holon what works and should be conserved; what does not function well and thus requires change. Microbes, plants, and animals are learning to alter and distribute genetic material in response to human attacks or gene implants.

SPECIAL CHARACTERISTICS OF HUMAN SYSTEMS:

Perceptual separation from other species: In the historic evolution of Western industrial culture, rooted in ancient Greece, Rome, and the Middle East and now spread to all parts of the world, there has been increasing focus on materialism and on linear verbal and technological communications. Western humans consciously opted for individualization—perceived separation from each other and from other species—with a gradual erosion of inner sensing and of inter- and intraspecies communications. The scientific perspective invalidates "inner" experience—dreams, mystic vision, telepathy, telempathy, precognition, remote sensing, etc. This historic process has cut us off from much natural interaction with other species and the ecosystemic, planetary, and cosmic holarchy in which we are embedded.

Except as they were brought into the Western system, indigenous and traditional peoples did not undergo this process, thus maintaining their natural "inner" knowledge and interspecies communications—all disrespected by Western science as "primitive."

Ethics and Law: The individuals of fish, bird, and mammal species appear to behave in accord with innate knowledge or "rules" of nonlethal interaction with same-species members, which ensures territorial sharing. Humans have apparently lost this evolved regulation of aggressive territorial (property) sharing. Therefore they must consciously design legal and ethical guidelines for behavior and for response to transgression of these guidelines. At present this process of setting rules has extended to global matters of resource sharing, trade, communications, etc.

Spirituality: Another characteristic of human systems is spirituality. Most of humanity acknowledges, pays tribute to, and is guided by spiritual concepts of Creation, Nature, Higher Power or Intelligence, and Cosmic Consciousness. Related beliefs and practices inspire awe and reverence, often forming the basis for ethics. Because spirituality is not acknowledged by science as a natural and life-promoting feature of humanity, it is largely concretized into religious forms that do not necessarily foster human health and harmony.

HARMAN:

That's really quite a fine list of common characteristics. At the moment I can't think of anything that I feel was left out.

CHAPTER
FIVE

Intelligence and Consciousness

Sooner or later a certain truth is brought home to you [namely, that consciousness] is the inner side of the whole, just as human consciousness is the inside of one human being. . . . Although it makes sense to inquire how and when consciousness developed into what we now experience as such, it makes no sense at all to inquire how and when mind emerged from matter. . . . Once you have realized that there is indeed only one world, though with both an inside and an outside to it, only one world experienced by our senses from without, and by our consciousness from within, it is no longer plausible to fantasize an immemorial single-track evolution of the outside world alone. It is no longer possible to separate evolution from evolution of consciousness.

—Owen Barfield (1982)

Not just animals are conscious, but every organic being, every autopoietic cell is conscious. In the simplest sense, consciousness is an awareness of the outside world... To live every organic being must sense and respond to its surrounds... Life is more impressive and less predictable than any thing whose nature can be accounted for solely by forces acting deterministically.

—Lynn Margulis and Dorion Sagan (1995)

HARMAN:

The present discussions on the epistemology required for accommodating the scientific exploration of consciousness seem to me to be only the beginning of an entirely new dialogue. It's like the Emperor's New Clothes: Once it's pointed out, everyone can see that ruling out consciousness as a causal factor has distorted the scientific picture of reality, until we really don't know what we "know for sure." Admitting an extended epistemology changes the rules of the game, and it's anything but clear what effect this will have on the biological sciences.

The epistemological and ontological assumptions that emerge out of these new considerations are revolutionary in their import. If indeed the scientific community attempts to construct a true science of consciousness using an appropriate epistemology, serious attention would have to be paid to the inner explorations that have gone on for thousands of years within the world's spiritual traditions. The distillation of these explorations has sometimes been termed the "perennial wisdom." In this ontology, consciousness precedes form, and creation is continuous.

Nobel laureate biologist George Wald may have had something like this in mind when he proposed (1987) that the major puzzles in the evolutionary picture, to his mind, could only be satisfactorily resolved by assuming that creative mind is not an emergent quality appearing only in the latter stages of the evolutionary process; creative mind appears to have been present all along, even before the first life forms. If we follow through with this kind of logic and ontological assumption, the story of evolution has a very different meaning from the accepted version. This is not to claim that this alternative view is "right"; only that some of the philosophical and biological puzzles appear in it with a different and perhaps more tractable form.

SAHTOURIS:

Let's try to come to some agreement on the broad issue of intelligence and consciousness. We have already shown that by conceptualizing and modeling the entire universe as a holarchy containing smaller holons in continual co-creation, its intelligent and conscious evolution makes far more sense than within the old mechanical model of a nonliving universe, which has had no explanation for its awesomely intelligent process since abandoning the *deus ex machina,* the inventor God outside the great mechanism, the Grand Engineer of the dualistic Cartesian universe.

We have indicated that we do not consider consciousness to be an emergent property of evolution-that we do not see how consciousness could emerge from nonconsciousness any more than intelligence can spring from non-intelligence or, for that matter, life from nonlife. I have shown consciousness and intelligence, in their simplest expression, to be logically inherent in autopoietic holons from the first. We have given repeated examples of biological intelligence, such as genetic change as an active response to a specific environmental impetus, and the strategy of prokaryotes building themselves into the cooperative communities we know as eukaryotes. But these are, after all, still reduction-ist approaches to the problem and amount to no more than descriptions of intelligent processes at these levels. The question is how to explain them.

Since we agree that holons in holarchies must be seen as aris-ing through the two-way interactions of the microscopic and the macroscopic, the smallest and largest "features" of the universe, let's look at intelligence and consciousness now at the most macroscopic levels.

By these macroscopic levels, I mean not only the greatest extension over space but whatever lies beyond all spatial exten-sion, beyond the physical, material world. Let me use "mind" as a shorthand term for intelligence and consciousness. We all have the experience of mind in our thinking, dreaming, and feeling; we are all aware that however real these experiences are, they are not localized in space and thus do not lend themselves easily to measurement. Some speak of them as being in other dimensions; we agree that nonphysical experience may better be described in levels. Thus the three dimensions of physical space comprise one level; all other reality, whether it is dimensional (physical) or not, is at other levels of experiential being.

The closest we can get to their scientific measurement is via their physiological frequency signal traces, recorded electronical-ly, as in the case of dreaming and thinking measured by the EEG. This is somewhat like studying particles by way of the traces their paths leave in bubble chambers. This is not the real thing, but at least an indication of its existence, with some hints about its nature. Such traces tell us nothing about dream content—just that dreaming "really" happens.

Valerie Hunt (1995) reports a large body of data on newly identified wave frequencies recorded from the body surface by radio telemetry developed for the space program. EKGs, EEGs, and EMGs all record frequencies between 0 and 250 hertz. Then there appears to be a gap up to about 500 hertz, and a whole new range of frequencies from 500 to 20,000 hertz, which correlate significantly with subjective mental and emotional events. Hunt also brought in people who claimed to see auras and found good correlations between their reports and measured changes in the patterns of these frequencies. What all this suggests is that we are more than our physical bodies; that we indeed have nonphysical or "subtle" aspects, known from time immemorial in other cultures, and which are drawing increasing attention in our own. Hunt herself calls them "mind fields," and her work is making them available for limited but useful biological and physiological study. Thus, improving our measurement equipment brings us a little closer to these other levels, just as in the earlier cases of smell, taste, and electromagnetism. One hopes that such evidence for the "reality" of subjective phenomena will make them more credible in their own experiential terms. It is interesting to note that Hunt's "mind fields" have no boundaries and are assumed to be co-extensive with the entire universe, making possible all sorts of effects at any distance.

Other levels of being have been investigated by theoretical physicists for some time, as concepts of higher dimensions or parallel universes. Michio Kaku (1994), for example, tells us the deeply ingrained prejudice that our world consists only of three spatial dimensions and one of time is "about to succumb to the progress of science." Eloquently, and at length, he explains that the laws of nature, when expressed mathematically under the assumption of a four-dimensional universe, are impossible to unify elegantly, while the elegance with which they can be expressed and united under the assumption of ten dimensions is extremely strong evidence for the existence of these other dimensions. However, he goes on to say that the mathematical evidence for these other dimensions is puzzling in that they appear to be so "tightly curled" that in reality they would be impenetrable or inaccessible. Physicists are actually suggesting, he says, that we may have to blast our way into them, though that would require more nuclear power than we can presently generate on Earth.

This is a very peculiar, if not also dangerous, view of things—the kind that comes about when a mathematical physics defines our world while discounting or ignoring phenomenological data. As I see it, Kaku is reporting the Western physicists' discovery via mathematics of what other cultures have always known through direct experience at other levels or realms of existence. What matters to our own development of a new biology is that these levels become scientifically credible. Perhaps these results in the world of physics will help open discussion of the experiential data of many cultures over millennia of time, which have already attracted such physicists as David Bohm, David Peat, Fritjof Capra, and Fred Allen Wolf. The traditional cultural data all seem to agree that at nonphysical levels consciousness is experienced as primary cause, whether it manifests materially or not, and this makes all these levels accessible through personal versions of consciousness or "mind fields."

Training to function competently at these nonphysical levels of being is often, in other cultures, as long and technical as the higher educational training of Western culture in functioning competently in specific ways within physical dimensions. Our military, and CIA, has discovered this, as recently revealed in its espionage training programs for "scientific remote viewing" (McMoneagle, 1993; Brown, 1996; Puthoff, 1996)—a technical procedure for doing what native and traditional cultures have long done in their own ways.

Multilevel reality, however new to Western science, was the standard worldview in the sophisticated ancient cultures of Taoist China, Vedic India, Greece, Western Africa, and the Mayan and Incan Americas. It also prevails among the indigenous tribal cultures of Australia, Indonesia, and the Americas—in short, in many cultures of all continents throughout history. Yet our labeling of all traditional cultures as "prescientific," which I will argue is indefensible, has kept us ignorant of vast bodies of information written off as superstition or as religious (that is, "imaginary") experience. Even Gary Zukav, who respects native knowledge, tells us that we are only now evolving the capacity to sense other levels of reality:

> "Since the origin of our species, we have evolved through the exploration of physical reality.... Until now our species has been limited in its perception to the five senses ... but the human species is now moving beyond these limitations ... in a new evolutionary leap ... the emergence of the multisensory human.... The singularly new perception of the multisensory human is this: SPIRIT IS REAL. The recognition, acceptance and inquiry into the nature of existence and intelligence that is both real and nonphysical is the foundation of the science that is now longing to be born."
>
> —Zukav (1991)

It was born long ago, and it has been the standard worldview of most historic cultures, not through their ignorance, but through valid perceptions we alone have denied, excluded from experience, and thus lost. Our task is to relearn them and to integrate their data with our Western knowledge and revise our epistemology accordingly.

It seems to me far more elegant to see consciousness as the basic "energy" of the universe, for lack of a better metaphor— energy that transforms itself from its "pure" or nonmaterial state to its material state, by way of the intermediate phase we call elecromagnetic energy. Consciousness is never lost in these transformations, as we know energy is not lost in converting to matter. It simply transforms itself from one state to another, with matter as a sort of end state, though still reversible. All ancient esoteric traditions tell us the error we make is thinking matter to be all there is. We now know there is electromagnetic energy; acknowledging its background or source as conscious- ness is the next step. Then everything in this magnificent self- creating universe falls into place as an intelligent improvisational dance—a cosmic experiment in doing whatever is possible within the constraints of the material state. I believe this basic sea of consciousness "energy" is what, in the history of science, has variously been called the plenum, the ether, the implicate order, and most recently, the zero-point energy field, though I don't think we need physics to validate the nonphysi- cal. In religions it is called spirit, Brahman, or by the different names of God. To personify it is a human choice that neither validates not invalidates what simply is.

The dynamic dance of nature is ever conscious at every level, from the tiniest particle to whatever its currently largest configu- ration, or holon, is. That is my basic assumption about the living universe, no stranger than any of the assumptions of physics. It is shared by all the indigenous cultures I have come in contact with, as well as all esoteric traditions. In fact, native peoples actively practice conscious communication with all other beings and aspects of nature, while our industrial society has cut itself off from that participatory universe. I believe that is why many scientists are now drawn to these cultures. Our task here is to translate this worldview of an intelligent conscious universe into a recognizable holistic biology.

HARMAN:

I think you're quite right about the tendency to use modern physics to validate the nonphysical. I'm very wary of the facile arguments, based on modern theoretical physics, that purport to make plausible the existence of other realms of existence. Physics is not based on the ontological stance that assumes a holarchic universe; there is no reason to assume that its extension comprises the best way to explore the nonphysical. But let me comment further on that in a moment.

Some of what you say is reminiscent of John Davidson's "Formative Mind" concept (1992). He speaks of three levels of mind: "conscious" mind, subconscious mind, and "formative" mind. In our normal lives it is not possible for us to know how or why things happen as they do. Sometimes the coincidence of events is such that we intuitively know there has to be a hidden connection. We may call it synchronicity, serendipity, luck. But what seems like chance to the conscious mind is understood quite differently if we become aware of the connections in the subconscious mind—or the formative mind.

In this ordinary world we all appear to have individual, separate minds. But at deeper or formative levels, all our minds are ineluctably linked. At the level of the "higher Self" we know ourselves to be "parts" of one great whole. At this level we understand ourselves to be co-creators or shareholders of our mind-dream in this physical world.

Interestingly enough, this concept that at a level beyond ordinary consciousness we know ourselves as co-creators of the world around us is being reconsidered within Western theology, where it certainly was not welcome at an earlier time. In a remarkable synthesis of science and Christian theology, Philip Hefner (1993) writes of the human being as "God's created co-creator."

A tacit corollary to the Darwinian thesis that the human form has evolved from animal forms is that human consciousness has biologically emerged from animal consciousness. Owen Barfield (1982) comes to a different point of view, expressed in the epigraph at the beginning of this chapter.

This sounds similar to Wald's conclusion that "Consciousness was there all the time." Likewise, Rupert Sheldrake's concepts of formative causation and "morphic fields," which most biological scientists consider preposterous, look much more plausible in this context. Robert Wesson (1990) describes a "third approach" to evolution besides creationism and neo-Darwinism. "Nontheistic nonmaterialism . . . lies midway between . . . the belief that the world is the work of a great personality who watches over it and perhaps intervenes on occasion to set things right; and the theory that material particles are the totality of existence." He postulates a *metacosmos,* a ground of being, that underlies the material world and makes it possible. "The material world is not the totality of existence but a derivative of something primordial."

Sheldrake, in the final chapter of his controversial *A New Science of Life* (1981) postulates that the physical universe is created, in an ultimate sense, by a transcendent consciousness. This transcendent consciousness is not developing toward a goal; it is its own goal. It is not striving toward a final form, it is complete in itself. Since this transcendent consciousness is the source of the universe and of everything within it, all created things in some sense participate in its nature. The more or less limited wholeness of organisms at all levels of complexity can be seen as a reflection of the transcendent unity on which they depend, and from which they are ultimately derived.

If we allow ourselves to think along these lines, it has many biological implications. First of all, we will think anew of the holarchy from molecules to organelles→cells→tissues→organs→organisms→societies (human and nonhuman)—to the human race or all biota. The organism has consciousness in some sense—at least we know that we do. C. G. Jung's "collective unconscious" is an assertion that societies and the entire human race in some sense share an aspect of consciousness. At the same time, since all is contained within the transcendent unity, we cannot rule out that an organ may have consciousness in some sense, or even a cell. In fact, when we consider the functioning of the body's immune system, it is hard to resist the temptation to think of lymphocytes' recognition of invading pathogens as representing some sort of consciousness. What Bergson terms an *élan vital* would seem to correspond to consciousness at the level of life as a whole. When some speak of Gaia as having consciousness, this is at the level of the earth as a whole.

Assuming such a hierarchy of conscious agencies, those at higher levels express their creativity through those at lower levels. One can imagine that if such a higher-level creative agency acted through human consciousness, the thoughts and actions to which it gave rise might be experienced as coming from an external source. This indeed is often reported in the phenomena of *inspiration*. Moreover, if such "higher selves" are immanent in nature, it is conceivable that under certain conditions, humans might become directly aware that they are embraced or included within them—and this, too, has often been described. Indeed, one can imagine that on occasion they might also be experienced as separate from the individual human (somewhat as in one's own mind, a portion may be experienced as separate—as one's conscience, for example, or as a demonic aspect). Again, this has been described as encounters with *devas* or nature spirits.

If physical matter is inseparable from the single unity, as is consciousness, then there is no "mind/body" problem. That is, there is no conceptual problem with the consciousness aspect of the human organism interacting with the material aspect. Nor is there a problem with the amazing instinctual abilities of "lower" organisms—and, indeed, with the "creative problem solving" abilities of the remarkable prokaryotes—since these could be seen as manifestations of a higher mind at, say, the species level.

The emergence of life on Earth is not as discontinuous an event as it seems in the conventional view, because in a sense there is nothing that is not alive. As the emergence of life on Earth can be viewed as a manifestation of creative mind, so too can the creative experimentation of the Cambrian explosion. They seem more understandable if one is allowed to hypothesize, with George Wald, that "consciousness was present all along."

Teilhard de Chardin postulated something of the sort in *The Phenomenon of Man* (1959, pp. 149-152):

> We now perceive a new way of explaining, over and above the main stream of biological evolution, the progress and particular disposition of its various phyla. It is one thing to notice that in a given line in the animal kingdom, limbs become solipedal or teeth carnivorous, and quite another to guess how this tendency was produced. It is all very well to say that a mutation occurs at the point where the stem leaves the verticil. But what then? The later modifications of the phylum are as a rule so gradual, and so stable are sometimes the organs affected, even from

the embryo, that we are definitely forced to abandon the idea of explaining every case simply as the survival of the fittest, or as a mechanical adaptation to environment. . . . The more often I come across this problem and the longer I pore over it, the more firmly is it impressed upon me that in fact we are confronted with an effect not of external forces but of psychology. According to current thought, an animal develops its carnivorous instincts *because* its molars become cutting and its claws sharp. Should we not turn the proposition around? In other words, if the tiger elongates its fangs and sharpens its claws is it not rather because, following its line of descent, it receives, develops, and hands on the "soul of a carnivore"? . . . To write the true natural history of the world, we should need to be able to follow it from *within*. It would thus appear no longer as an interlocking succession of structural types replacing one another, but as an ascension of inner sap spreading out in a forest of consolidated instincts. Right at its base, the living world is constituted by consciousness clothed in flesh and bone. From the biosphere to the species is nothing but an immense ramification of psychism seeking for itself through different forms.

As it was a long path from Copernicus and Galileo to the understandings of present-day life sciences, so there is much to be done before we can say we understand biology anew within a new cosmology. The invitation posed by the hypothesized "holistic revolution" in the biological sciences is that of freeing us to consider all of the mysteries of nature without, perhaps unconsciously, sorting out those which are conceptually acceptable from those that fall outside the bounds of the comprehensible as conventionally understood.

SAHTOURIS:

You are right, of course, to point out that some of our leading Western scientists have been talking about the universal intelligence, or mind, of the material world for some time, despite the "official" view that such notions are heretical. In his foreword to L. J. Henderson's *The Fitness of the Environment,* Wald wrote "A physicist is the atom's way of knowing about atoms." Quoting this in his article on "The Cosmology of Life and Mind" (1987), he continues, arguing that "The stuff of this universe is ultimately mind-stuff. What we recognize as the material universe, the universe of space and time and elementary particles and energies, is then an avatar, the materialization of primal mind. In that sense, there is no waiting for consciousness to arise. It is there always."

Wald points out that Sir Arthur Eddington had said, already in 1928, that "the stuff of the world is mind-stuff"; that Wolfgang Pauli suggested in 1952 that "physis and psyche (that is, matter and mind) could be seen as complementary aspects of the same reality"; and that von Weizacker in 1971 added that "consciousness and matter are different aspects of the same reality." I know you don't think that physics has much deep insight to offer biology; however, these highly respected Western scientists can be seen as offering biologists a primary role in science by acknowledging an intelligent, therefore alive, universe. I think this is how Wald sees it and why he cites them. Unfortunately, few biologists have leaped at this wonderful opportunity, though Erich Jantsch defines mind as the self-organizing dynamics of a system, and Gregory Bateson said that epistemology itself is the integrated metascience of mind, which has evolution, thought, adaptation, embryology, and genetics as its subject matter.

Let me refer back to that Bergsonian notion of contacting reality in two ways, through physical senses and through deep intuition, which is similar to Barfield's perception of the outside world through our senses and the inside world through consciousness. I have myself written about the inner and outer "ways of knowing": one given primacy in the East, the other given primacy in the West, while native cultures often seem to use both with equal ease. My growing familiarity with Andean cultures— and the Incas were highly scientific—now makes me see these very distinctions I used as unnecessarily dualistic. In the Andes, what Barfield calls the inner and outer world are no more different than up and down are different, or right and left. Both are equally real and both are perceived by senses and by consciousness, rather than by one or the other. Children are taught to touch stones and to speak with them, they are not taught that anything in their experience, day or night, is unreal or imaginary, though they are taught very strictly not to lie, not to distort experience to deceive others. That is, truth and untruth are categories of experience; reality and nonreality are not.

Do we not hear and see in what we call the "inner" world as much as in the "outer" world? Are we not conscious in both? Is one any more real than the other? Is there really a significant difference between them, except that in our culture we are more comfortable in the "denser" version of reality and so write off the rest as somehow "imaginary" or "unreal"? We learn to call them inner and outer, but these are socially determined categories.

Consider the shared experience of countless people throughout history in an "out-of-body state." In my own experience of that state, which in our culture requires training, I perceive myself located within what we call an "inner" world, yet it is as physically "solid" as the "outer" world, although when I perceive the outer world from the perspective of this inner world, the outer world is nonmaterial-that is, I can pass through its walls, while I cannot pass through the walls of the inner world. For this reason, I hesitate to call the inner reality nonmaterial. Even physicality is relative to one's perspective. If we perceive mind as boundless, it is useless to argue that such experiences are "only in the mind." In any case, I think we want to work hard to eliminate dualities that make one aspect of experience more valid than another, or one reality more real or more physical than the other. This brings up the Vedic view that what we believe to be reality may be the dream, though that, of course, is dualism turned around.

Like Wald's and Sheldrake's, and I believe yours, my own current and still evolving perspective shows consciousness as primary. Thinking as nondualistically as I can, physical worlds come about through the continual autopoietic manifestation of consciousness into diverse patterns. Consciousness, that is, evolves through its endless self-creations and is indistinguishable from them. Spirit is Nature and Nature is Spirit, if you like those terms, though the concepts, not the labels, are important here. Now these manifestations of consciousness as living systems may or may not be physical or material from our human perspective. They are simply, to use my favorite metaphor once again, the great improvisational dance of nature at all its frequency levels. And looking into the dance, as in looking into the atom, we will ultimately find no substantive dancer, only the energetic dance itself. That is the magic of reality as we are now coming to terms with it, and the reason we need new metaphors.

We simply have to recognize that modern science—Western science—defined reality in a very limited way, abstracting it as a particular physical world experienced through a few counted physical (biological) senses, ignoring all other aspects of experience, even of that limited experience in the sense that our bodies could not function only at the frequency levels accepted by science as real. These limits were set to make things appear to be as understandable as the mechanisms we were inventing at the time science was founded—hence Descartes' insistence that God's nightingale was exactly like the wind-up kind, only more complex.

It was a way of convincing ourselves that we could understand and control all of that mysterious, unfathomable, feminine, threateningly-out-of-male-control "nature." Brian Easlea (1983) pointed out that Francis Bacon, as the father of the scientific method and a frequent attendant at witch trials, chided his fellow scientists as schoolboys expecting nature to unveil herself at their request, while urging them instead, like grown men, to pursue, hound, and torture her to reveal her secrets. We simply cannot afford to sweep these origins of science under the rug any longer; there is too much at stake. This manly science was quickly put into the service of a developing technological and manufacturing elite who claimed they would solve all human problems, and this is where science, now somewhat desperately in search for technological solutions to the ever greater problems caused by technology, remains today. We need to rethink the whole enterprise of Western science from a more globally historic perspective, to reconsider its assumptions and its practice, to recognize the need for profound changes that can put science in service to humanity as a whole.

HARMAN:

Yes. Although what you propose takes a bit of humility. We moderns are reluctant to recognize Western science as an artifact of European culture. But there really is no valid reason to suppose that reductionistic science by and of itself can ever provide an adequate understanding of the whole.

What the "holistic revolution" in biology comprises, as I see it, is to retain the ideals of the open-minded scientific spirit and the tradition of public validation of knowledge (repudiating any scientific priesthood), but to open up the field of inquiry to the entire holarchy and to nonphysical as well as physically measurable aspects. Whether that will be soon done within science is an open question. However, because of the cultural shift that appears to be taking place, which attaches increasing importance to the transcendental, there may be increasing public insistence that some such development take place in science if science is to retain its present position as the *only* generally accepted cognitive authority in the modern world.

SAHTOURIS:

There is a real problem in that Western culture deeply assumes its superiority over all other cultures and that Western science has set itself up as the *only* science, the only arbiter of what is real and what constitutes a valid understanding of reality. Before I show that other cultures *have* valid sciences, by definition, let me insert two quotes here to show how Western culture and then its science look from the perspective of another culture.

Nicolas Aguilar Sayritupac, an Andean Aymara Indian, wrote, in a prologue to Carlos Milla Villena's book *Genesis de la Cultura Andina* (1983) (my translation from the Spanish, itself a translation from Aymara) :

> From Chucuito, my village, at the edge of the sacred lake, the nights of May are beautiful and dazzling. As it has always been, our elders showed us the Southern Cross, with

its two guiding stars, as four small suns that guide our community and our thought through the black nights and passages in which our happiness and faith sometimes fall. I am old now and too tired to walk strange roads. Many times I have lost tranquility and hope when so often I saw my community and my culture crushed and destroyed. . . . In the hearts of the Westerners there are no feelings that can resolve their conflicts; their hearts go in one direction and their minds in another. It follows that men, women, children, and old people do not work together collectively. . . . The human being of the West has abandoned being human and has turned himself into an individual: man, woman, child, elder, separate; community has died in them, the ayllu-the essential unity of humanity. The existence of Western people and society has been destroyed by their egoism. . . . On the contrary, we Indians have things well in our heads, our feelings in order, determined to do what we can; it is for this reason that we do not go away much from our home and family, for this reason that we have kept ourselves away from the equivocal ideas of the men who find themselves in the place where the Sun hides itself. . . . I came back to see in May, and in all the nights of the year, the four beautiful brilliant stars and their two star guides, as when in the nights many years ago, my father, looking with his good eyes, said to me: Look at the *Chacana*. . . now be certain that if the West wants to totally destroy our community and our culture it will first have to destroy the Cross of May in the heavens. The Aymaras are eternal people.

This critique of Western culture, which I find both appropriate and poignant, is relevant to my earlier discussion of the loss we sustain in our extreme individuality and our denial of nature's intelligent process and human connection with it at all levels. Now let us look at the critique of Western science in the introduction to the same book, where Salvador Palomino F. and Javier Lajo write:

> Does Andean science exist? It has always been considered in the circles that manage Western culture that other cultures either have not had any scientific development or that their systematized knowledge has been so poor that it was easily assimilated into the hegemonic world culture. Thus it is in America, particularly in the Andean region, which in ancient times was based in Tawantinsuyo. Among the principal causes of this undervaluation of the non-Western, we can identify ignorance of other cultures, but there also exists, without a shadow of a doubt, an accumulation of prejudices which, as principles or dogmas, the West uses to preserve its hegemony—principles that when applied to the study of occidental reality explain it objectively, but when applied to non-Western realities degenerate into norms or molds whose narrow frameworks attempt vainly to repress objective phenomena or knowledge that escape the rationality and the methods of the West. What these principles cannot make simple, is denied, silenced, or disqualified as obscurantism, esotericism, charlatanism or witchcraft. (My translation from Spanish.)

I have read equally appropriate critiques of Western science from the perspective of Arab and Vedic scientists, and I believe we should be open to such views in order to expand our own perspective on human knowledge as well as that knowledge itself. "Science" is defined by *Merriam Webster's Collegiate Dictionary*

(Tenth Edition, 1993) as "the state of knowing" or "a department of systematized knowledge as an object of study." The *American Heritage Unabridged Dictionary of the English Language* (Third Edition, 1992) defines science as "the observation, identification, description, experimental investigation and theoretical explanation of phenomena." A bit more precise, yet a good description of what non-Western cultures have done that is appropriately dignified with the label "science." As defined by the *Oxford English Dictionary,* science is "the state of knowing" or "knowledge as opposed to belief or opinion"—knowledge that is "acquired by study." The OED continues explaining that science is "in a more restricted sense: a branch of study which is concerned either with a connected body of demonstrated truths or with observed facts systematically classified and more or less colligated by being brought under general laws, and which include trustworthy methods for the discovery of new truth within its own domain."

Detailed as this definition is, there is nothing in it to exclude indigenous and other non-Western sciences. It is quite unreasonable for Western scientists to limit not only their reality or worldview but science itself to its Western practice. Other perspectives and painstakingly researched phenomena can greatly enrich our own scientific knowledge and keep us from having, in many cases, to reinvent the proverbial wheel. So let us proceed to discuss other cultures with due respect for their sciences and greater impartiality in judging which cultures' sciences have the best explanatory power for describing the great variety of human experience.

Some of the scientists who do see conscious intelligence at the heart of nature have discovered it through their own inner explorations, through the perennial philosophy, or through both personal and intellectual exploration. Others, like myself, have also discovered it by association with native people. Ecologist David Abram (1996, p. 10), for example, associated himself with traditional medicine people in Indonesia and Nepal, living with them for some time. He describes his shaman associates as living at the edges of their communities, maintaining the balance between the community as a whole and its surrounding ecosystem, as well as healing individual illness. He defines their abilities as:

> . . . heightened receptivity to the meaningful solicitations—songs, cries, gestures—of the larger, more-than-human field . . . the experience of existing in a world made up of multiple intelligences, the intuition that every form one perceives—from the swallow overhead to the fly on a blade of grass itself—is an experiencing form, an entity with its own predilections and sensations, albeit sensations that are very different from our own. . . . It is not by sending his awareness out beyond the natural world that the shaman makes contact with the purveyors of life and health, nor by journeying into his psyche; rather, it is by propelling his awareness laterally, outward into the depths of a landscape at once both sensual and psychological, the living dream that we share with the soaring hawk, the spider, and the stone silently sprouting lichens on its coarse surface.

The Western dualistic misunderstanding of this communion with other aspects of living nature is further elaborated, as Abram warns us, that when witnessing the shaman :

> . . . entering trance and sending his awareness into other dimensions in search of insight and aid . . . we should not be so ready to interpret these dimensions as "supernatural," nor to view them as realms entirely

"internal" to the personal psyche of the practitioner. For it is likely that the "inner world" of our Western psychological experience, like the supernatural heaven of Christian belief, originates in the loss of our ancestral reciprocity with the animate earth. When the animate powers that surround us are suddenly construed as having less significance than ourselves, when the generative Earth is abruptly defined as a determinate object devoid of its own sensations and feelings, then the sense of a wild and multiplicitous otherness (in relation to which human existence has always oriented itself) must migrate, either into a supersensory heaven beyond the natural world, or else into the human skull itself—the only allowable refuge, in this world, for what is ineffable and unfathomable.

When Abram returned from Indonesia and Nepal, where he had learned this broader perception of, and communication with, all living entities in nature, he found that squirrels "swiftly climbed down the trunks of trees and across lawns to banter with me" and experienced becoming one, for hours at a time, with herons fishing. But then he goes on to say:

> Yet, gradually, I began to lose my sense of the animals' own awareness. . . . I now found myself observing the heron from outside its world, noting with interest its careful high-stepping walk and the sudden dart of its beak into the water, but no longer feeling its tense yet poised alertness with my own muscles. . . . And, strangely, the suburban squirrels no longer responded to my chittering calls. Although I wished to, I could no longer focus my awareness on engaging their world as I had so easily done

a few weeks earlier, for my attention was quickly deflected by internal, verbal deliberations of one sort of another-by a conversation I now seemed to carry on entirely within myself. The squirrels had no part in this conversation.

He lists the beliefs of our culture—that other species are not as awake and aware as humans, that they have no language, that their behaviors are coded within their physiologies—saying "the more I spoke about other animals, the less possible it became to speak to them." Most of Abram's book is devoted to understanding how our Western culture erected the barriers between ourselves and the rest of nature that prevent communion with it.

I myself have shared Abram's painful experience of losing communion with the other beings of nature each time I have returned from time spent with indigenous people on their turf. It is not only the pain of losing that communion, but also the pain of being unable to convey these entirely natural interactions with plants and animals that sound like fantasies to Western colleagues. How can one adequately communicate one's experiential knowledge that indigenous science is as valid as our own and indeed far wiser?

HARMAN:

That's a very moving description of our predicament. My own experience, though of a somewhat different sort than yours, checks with it completely.

I do want to mention two areas of scientific research (besides cultural anthropology) that aim in a supportive direction: research on nonordinary states of consciousness and research relating to "meaningful coincidences" and "paranormal" phenomena. (See the "Interlude" following this chapter.)

SAHTOURIS:

The holarchic, living-system paradigm sheds a great deal of light on our "predicament." All organisms, from micro-organisms to the total environment and the Earth itself, are all evolving together with intelligence exchanged at all levels. Some of my insights on this have come, as I said earlier, from outside the Western scientific paradigm, from various indigenous scientists. Their experience in integrating with nature in order to study it, rather than isolating pieces of nature in laboratories to control them, is key to understanding Earth-and-ourselves as a single whole system. Western science has been instrumental in developing technologies, but in the course of this development we are destroying the natural context on which our survival depends. Thus Western science is currently failing us as a science for survival. It was this need for a healthier science of survival that led me to learn more about indigenous science, and I hope we can bring its understanding of consciousness as inherent in all nature as well as its ecological understanding and wisdom into a broader science for all humankind.

HARMAN:

Yes, it's most interesting that recognizing the need for a new epistemology for a "science of consciousness" leads to some of the same places as does respect for the experience of the indigenous peoples. Of course, if we really take seriously the "radical unity" metaphor, and recognize deep intuition as a possible probe into the nature of reality, neo-Darwinism takes on a new form. The concept of "formative mind" totally changes the picture.

A dominant myth of modern society, coming out of a reading of Western science, is (a) that the essential characteristics of human nature are to be understood as the consequence of an evolutionary succession of random events (from the origin of life to later mutations) and natural selections, and hence accidental-without purpose or meaning; (b) that the essence of ourselves is thus to be found in the DNA with which we were born; and (c) that mind or consciousness emerged near the end of this long evolutionary process and is to be understood in terms of its physiological origins. This myth infuses our education, health care policy, legal justice system, and other social institutions. If it were found to be fundamentally in error, the implications are far-reaching.

The chief reason to suspect that this "central myth" may be in error is that in science's exclusion of consciousness as causal (by an ontological assumption made early in the history of science), a fundamental bias was introduced that is far more basic than the "Newtonian" bias prior to the advent of modern physics. That exclusion forces biological scientists to seek mechanistic explanations, even in situations where those explanations seem obviously inadequate. (A typical example is the insistence that all the information and motive force to guide ontogeny from fertilized egg to adult organism must reside in some as-yet-undiscovered "program" in the DNA.)

It would have been impossible to forecast the characteristics of modern society from the shift in metaphysical premises we know as the scientific revolution. Similarly, we cannot expect to predict the societal implications of a change in our "official" stories of evolution, ontogeny, morphogenesis, heredity, and so on. We may safely anticipate that they will be profound.

For example, consider the impact of a changed view of physical death. If indeed we humans fundamentally evolved by mechanistic processes out of a material universe, and if life is basically a set of very complex physical and chemical processes regulated by coded messages in the DNA, then when those processes stop we die, and that is the end of us—as physical organisms, surely, but in every other sense as well. If our consciousness, our cherished understandings and values, our individuality, our personhood, are simply creations of those processes, then when those processes stop we are no more. That is surely a fate to be feared, and indeed the fear of death permeates our society, disguised in a multitude of ways in which we seek "security," manifesting in the form of numerous other fears, driving us to seek solace in acquisitive materialism, and causing us to expend a disproportionate amount of our medical care resources keeping physical bodies alive long after the point at which there is any hope of further fruitful life.

However, the core wisdom of Western tradition—along with essentially all other traditions—has disagreed decisively with the above conclusion. The cherished values of Western tradition are based in the assertion that we exist in an essentially meaningful universe in which the death of the physical body is but a prelude to something else. Modern biological science dismisses this as a prescientific notion and wishful thinking. But perhaps this judgment will appear differently if the "holistic revolution" indeed takes place.

This issue of consciousness and survival is only one aspect of the shifting worldview; it is a useful one to explore, however, because it illuminates the strength of our prejudices. Serious attempts have been made to explore the concept of the continuation of personhood after physical death, and the evidence gathered has been disturbing to both positivist scientists and convinced religionists, because it fails to conform to their preconceptions. However, if that evidence is explored with humility and open-mindedness, it seems to point to features of an emerging "new myth."

One can imagine how much of the fear in our society would disappear if a new view of death were to become real in our lives—if we came to realize that we couldn't nonexist if we wanted to.

SAHTOURIS:

That is certainly among the fascinating possibilities opened up by the new paradigm we are proposing. It all seems to point to the conclusion that a holarchic biology which includes consciousness as its most fundamental feature simply makes more sense—more common sense and more sense of the available data—than our nonconscious mechanism metaphor for nature. This is how I see the shift from dumb mechanics to intelligent organics.

When we look again at our current story of evolution in this new framework, we will see that nature's improvisational dance of evolution is filled with experiential lessons for our own troubled species at this important time in its own historical evolution. For example, the story of the amazing prokaryotes related above or the lessons of politics and economics inherent in the evolution of our bodies.

HARMAN:

I think you're quite correct in observing that profound change in our biology-based "story" of our origins and basic nature and profound change in our social institutions will happen together. But before we go on to explore this interaction between biology and society, I'd like to briefly push even deeper regarding the power and limitations of modern science. My conviction continues to grow that the technological prowess and awesome complexity of modern reductionistic science has been allowed to conceal from our view the mischief caused by its presumed "monopoly" with regard to truth.

A basic problem with Western science has been long recognized but not widely discussed. Two philosophers, Pierre Duhem and W. V. O. Quine, one French and the other American, decades ago advanced one of the most important propositions in the history and philosophy of science. Basically, the Quine-Duhem thesis (Quine, 1960) is that any scientific hypothesis is embedded in a theoretical network that involves assumptions implicit in "observations," auxiliary assumptions, related hypotheses, "basic laws," the accepted nature of scientific methodology, common sense beliefs, and so on. When observations fail to confirm a hypothesis, it means that *somewhere* in that network there is a falsity. There is no way to tell just where in the theoretical network the falsity lies. Thus in the face of persistent anomalies and enigmas consideration must be given to revising any or all of the elements of the network.

According to the Quine-Duhem thesis we must give up the idea that we can use experience *either to confirm or to falsify* particular scientific hypotheses. Evidence does not itself determine our evaluation of hypotheses. When experience contradicts science, the science must be changed, but there are a number of different ways this might be attempted. There is no unique logic for determining exactly what to change in one's theory: Any hypothesis can always be "protected" and the falsity shifted to other statements elsewhere in the theoretical network. Thus it is possible for competent and rational scientists to disagree even after a great deal of data has been accumulated (as they clearly do, for example, with regard to the best way to explore life on Earth).

Reality is much too rich to be fully represented in the sorts of models and metaphors that are the stuff of science. Thus good scientists tend to ask not is a theory "proven" but "does it adequately represent the phenomena for specified purposes?" A consequence of the Quine-Duhem thesis is that *even our epistemological convictions* about what it is to acquire knowledge and about the nature of explanation, justification, and confirmation—about the nature of the scientific enterprise itself—may be subject to revision and correction. It is precisely to that point which the many scientific paradoxes associated with intentionality, teleology, consciousness, and so on seem to have brought us.

Modern Western science fundamentally entails three important metaphysical assumptions:

*a. **Realism*** (ontological—leads to epistemological conclusion). There is a real world that is, in essence, physically measurable (positivism). We are embedded in that world, follow its laws, and have evolved from an ancient origin. Mind or consciousness evolved within that world; the world pre-existed before its appearance, and continues to exist and persist independent of consciousness.

*b. **Objectivism*** (epistemological and ontological). This real world exists independently of mind and can be studied as object. That is, it is accessible to sense perception and can be intersubjectively observed and validated.

*c. **Reductionism*** (epistemological). This real world is described by the basic laws of physics, which apply everywhere. The essence of the scientific endeavor is to provide explanations for complex phenomena in terms of the characteristics of, and interactions among, their component parts.

These underlying assumptions are directly challenged by a wide range of biological puzzles, "anomalous" phenomena, and human experience. It is these metaphysical foundations of modern biology that are challenged by a holistic epistemology. Yet the holistic approach in no way invalidates the science we have; it puts it in an expanded context.

Interlude Three

Aspects of Research on Consciousness

If something like the holistic epistemology discussed in Chapter One were to be adopted as the scientific community attempts to construct a true science of consciousness, it would naturally follow that serious attention would have to be paid to the "inner research" that has been fostered for thousands of years within the world's spiritual traditions. This is as truly research as that carried out in the most modern scientific laboratory; the difference is in the accepted epistemological assumptions.

Whereas modern science is founded on such epistemological assumptions as objectivism, positivism, and reductionism, what has been called "transpersonal" research assumes the primacy of experience. Its fundamental epistemological assumption was formulated by Henri Bergson in his "philosophy of process." This assumption is that there are two ways in which we contact reality: through the physical senses (leading to empirical science) and through the deep intuition (the route of mystical philosophy). A complete science would acknowledge and employ both. The importance of this issue shows up in a central ontological question, namely whether consciousness is *caused* (by physiological processes in the brain, which in turn are consequences of the long evolutionary process) or is *causal* (in the sense that consciousness is not only a causal factor in present phenomena but also a causal factor throughout the entire evolutionary process). Western scientific method urges toward the former choice in both cases, whereas the phenomena of consciousness suggest the latter choice in both cases.

The distillation of the inner explorations of diverse spiritual traditions has been termed the "perennial philosophy"; it brings with it ontological implications, which are examined in a recent paper by Ken Wilber entitled "The Great Chain of Being" (1993). Based on some very sophisticated (if prescientific) exploration, this ancient view centers around the following proposition: "Reality, according to the perennial philosophy, is composed of different grades or levels, reaching from the lowest and most dense and least conscious to the highest and most subtle and most conscious. At one end of this continuum of being or spectrum of consciousness is what we in the West would call 'matter,' or the insentient and the non-conscious, and at the other end is 'spirit,' 'godhead,' or the 'superconscious' (which is also said to be the all-pervading ground of the entire sequence). . . .The central claim of the perennial philosophy is that *men and women can grow and develop (or evolve) all the way up the hierarchy to Spirit itself,* therein to realize a 'supreme identity' with Godhead."

A central understanding of this "perennial wisdom" is that the world of material things is somehow embedded in a *living* universe, which in turn is within a realm of consciousness, or Spirit. Similarly, a cell is within an organ, which is within a body, which is within a society . . . and so on. Things are not—cannot be—separate; everything is a part of this "great chain of being."

As Wilber observes, Western science became restricted to the matter end of the continuum only, and to "upward" causation only—causation from the material to the mental and spiritual; not the reverse. With that restriction came a faith that in the end, a nomothetic science could adequately represent reality—a faith that phenomena are governed by inviolable, quantified "scientific laws." From that restriction came both the power of modern science (basically, to create manipulative technology) and the limitation of its epistemology. From it also stem all sorts of classical "problems"—the "mind/body problem," "action at a distance," "free will versus determinism," "science versus spirit," etc.

This restriction of science to only a portion of "the great chain of being" was useful and justifiable for a particular period in history. The only mistake made was to become so impressed with the powers of prediction-and-control science that we were tempted to believe that that kind of science could lead us to an understanding of the whole. Fundamentally, *there is no reason to suppose that reductionistic science can ever provide an adequate understanding of the whole.*

What must be done now, according to Wilber, is to retain the open-minded scientific spirit and the tradition of open, public validation of knowledge (that is, abjuring any scientific priesthood), but to open up the field of inquiry to the entire continuum and to downward as well as upward causation. Whether the scientific community will soon choose to do that is a good question. However, because of the cultural shift that appears to be taking place—attaching increasing importance to the transcendental—public pressure may force such a change if science is to retain its present position as the only generally accepted cognitive authority in the modern world.

Indeed, there are at least two areas of research that point more or less in this direction. These are research on nonordinary states of consciousness and research relating to "intentionality" and "meaningful coincidences."

Research on nonordinary states of consciousness

One of the more accessible discussions of the aspects of conscious and unconscious mind is Roberto Assagioli's (1965). He uses the diagram below to include, coordinate, and arrange in an integral vision the data obtained through a multitude of reported observations and experiences:

1. The Lower Unconscious

2. The Middle Unconscious

3. The Higher Unconscious

    or Superconscious

4. The Field of Consciousness

5. The Conscious Self or "I"

6. The Higher Self

7. The Collective Unconscious

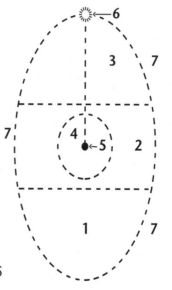

This is, admittedly, a crude and elementary picture. Nevertheless, it provides a helpful visualization for bringing out several features. The meaning of the various regions on the diagram is as follows:

1. *The Lower Unconscious.* This is similar to Freud's *id*—the psychic domain where sexual and aggressive drives and repressed experiences and memories are stored. The Lower Unconscious contains the elementary activities that direct the life of the body; the fundamental drives and primitive urges; various complexes, charged with intense emotion; dreams and imaginations of an inferior kind; lower, uncontrolled parapsychological processes; various phobias, obsessions, compulsive urges, etc.

2. *The Middle Unconscious.* The Middle Unconscious is formed of psychological elements similar to our waking consciousness and easily accessible to it; in this region our various experiences are assimilated and our ordinary mental and imaginative activities are elaborated and developed. It is similar to Freud's "preconscious," and contains recently forgotten and easily recalled items such as phone numbers of friends, memories of what we did last holiday, names of business associates, and so on.

3. *The Higher Unconscious or Superconscious.* From this level of the unconscious we receive our higher intuitions and inspirations—artistic, philosophical or scientific, ethical "imperatives" and urges to humanitarian and heroic action; the source of the higher feelings such as altruistic love, of genius, and of the higher psychic functions and spiritual energies. For most of us, most of the time, the contents of the Higher Unconscious are not readily accessible. It is a central focus of Transpersonal Psychology and the various spiritual disciplines to work at increasing this accessibility to repressed intuitive, creative, aesthetic, and spiritual potentialities.

4. The Field of Consciousness. This is that part of our personality of which we are directly aware, including sensations, images, thoughts, feelings, desires, impulses. This is our immediate awareness, the perceptions and mental events we are conscious of right now.

5. The Conscious Self or "I". The point of pure self-awareness, the center of consciousness; that which can be aware of, and attend to, areas in the field of consciousness. It is our "ego"—the outpost of the Self. It is Self experienced at the level of the individual, within the drama, trauma, joys and tribulations, hustle and bustle of our day-to-day activities. The "I" is the core observer that can witness and observe the drama of our lives.

6. The Higher Self. Here is the locus of the true Self, which remains when the conscious self seems to disappear through sleep, fainting, hypnosis, or narcosis;

related to the "hidden observer" in hypnosis research and the "inner self-helper" experienced in studies of multiple personality disorder. Conscious realization of the Higher Self or the transpersonal Self is sought through various yogas and meditative disciplines.

7. *The Collective Unconscious.* The outer line of the oval of the diagram should be regarded as delimiting the self but not disjoining it from the larger environment. It is analogous to the membrane delimiting a cell, which permits a constant and active interchange with the whole body to which the cell belongs; it allows a sort of "psychological osmosis" with other human beings and with the general psychic environment.

The above diagram helps to reconcile a seeming paradox with regard to the self. There seems to be a duality about the self: It is as if there were two selves—the personal self, generally unaware of the true Self, even to the point of denying its existence, with the latter latent, not in general revealing itself directly to our conscious awareness. However, the Self is one; it manifests in different degrees of awareness and self-realization. The reflection of the Self as the conscious self appears to be self-existent but has, in reality, no autonomous substantiality.

State of research

There has been a good deal of clinical research in psychotherapy and research in such areas as hypnosis, unconscious perception, selective attention, mental imagery, sleep, and dreams, all of which sheds considerable light on relationships between the self, the field of consciousness, and the middle unconscious. Carl Jung originated the term "collective unconscious," and he viewed his research on archetypes and on synchronicity as being major probes into that area. Since most scientists can see no possibility of a "mechanism" whereby communication with a collective unconscious or with other persons via a collective unconscious could take place, the concept remains in the far fringes of science.

The primary interest in altered states of consciousness research is in exploration of the higher unconscious (Harvard sociologist Pitirim Sorokin used the term "supraconscious" in his research on "creative altruism"). This is a central focus in "transpersonal psychology" (Bruce Scotton et al. [1996], Charles Tart [1975a and 1975b]), although it must be admitted there is no generally accepted research paradigm for this discipline.

The most extensive research on the higher unconscious, and the relationship to the ego and the Self, is in the spiritual disciplines. Roger Walsh and Frances Vaughan (1993) remind us of the broad range of possible altered states of consciousness (ASCs):

For example, the varieties of ASC that have been identified in Indian meditative and yogic practices alone include highly concentrated states, such as the yogic *samadhis* or Buddhist *jhanas;* witness-consciousness states in which equanimity is so strong that stimuli have little or no effect on the observer; and states where extremely refined inner stimuli become the objects of attention, such as the faint inner sounds of *shabd yoga.* Some practices lead to unitive states in which the sense of separation between self and world dissolves, such as in some Zen *satoris.* In other states all objects or phenomena disappear, such as in the Buddhist *nirvana* or Vedantic *nirvikalpa samadhi;* and in still others all phenomena are perceived as expressions or modifications of consciousness, such as *sahaj samadhi.*

[There is considerable agreement that it is] difficult to fully appreciate and comprehend altered states without direct experience of them. Indeed, such experiences can radically alter one's worldview and those who have them are particularly likely to regard consciousness as the primary constituent of reality.

Some of these spiritual traditions include very sophisticated epistemologies; more phenomenological than that of science, but no less disciplined. It is perhaps this area, more than any other, where fruitful exploration and wide acceptance of the results await a resolution of the question of an appropriate epistemology for scientific research in this area.

Research on "meaningful coincidences."

The term "meaningful coincidences" was first introduced by Jung (1955) and used with expanded definition by John Beloff (1977), to imply "acausal connectedness" between events separated by time or space. In plain language, it refers to events where there appears to be no possibility of physical connection in any known sense, and yet there is *meaningful* connection. Jung is very clear about the fact that meaningfulness is not to be interpreted as merely a subjective feeling. Synchronicity (his alternate term), he insists, "postulates a meaning which is *a priori* in relation to human consciousness and apparently exists outside man." At one point he suggests that synchronicity may be regarded as standing alongside space, time, causality, and energy as a fundamental dimension of objective reality.

As Beloff comments, "The possibility of a relationship that is at once acausal and yet meaningful presupposes a different cosmology." In fact, it was precisely the denial of such an acausal, yet meaningful, connection that justified the precipitous dumping into the rubbish heap of scientific progress of the three great occult arts of astrology, alchemy, and divination. Beloff prefers to relax the definition to include phenomena where the connection *appears* to be acausal, leaving it to further exploration to determine whether a kind of cause would seem to be present. In this expanded sense "meaningful" may refer either to the subjective judgment of the observer, or to a judgment based in historical data (as in the case of astrology or the I Ching).

As we are using it, then, the term "meaningful coincidences" includes Carl Jung's "synchronicity" (David Peat, 1987) and most of the range of the "paranormal." Examples include apparently "telepathic" communication, seemingly clairvoyant "remote viewing," and the "coincidence" between the act of prayer and the occurrence of the prayed for, such as healing. Another example is the feeling of having a "guardian angel" when a person feels warned about a danger or is provided with a particularly fortuitous circumstance in life. A host of historical and anecdotal examples fall into the categories of "miracles" and "psi phenomena."

The first broad attempt to explore this area systematically and in the scientific spirit would seem to be the work of the group around F. W. H. Myers at Cambridge, which resulted in the formation of the Society for Psychical Research in 1882. A helpful review of the field can be found in Mitchell, 1974.

The phenomena in this area fall rather naturally into three categories:

1. *Those that involve information alone.* In these cases, information is apparently obtained through means other than the known sensory channels. The information may come in waking consciousness, in a trance state, or in dreams. It includes *telepathic communication* (mind to mind); *clairvoyant perception* (remote viewing), *precognition* ("remembering" an event that hasn't happened yet), and *retrocognition*

("remembering" a past event that one has no knowledge of in the ordinary sense). While some attempts have been made to postulate some kind of mechanism for such instances, none has been successful. The phenomena appear to genuinely involve situations where there are two events—one, an awareness in the mind of the person receiving the information; the other, an event remote in space or time, or a thought or image in some other person's mind— which often have a profound meaningful connection to the persons involved, whereas there is no conceivable physical communication between the two.

Individual examples of these phenomena are often quite startling and convincing. Certain persons, sometimes called "psychics," reputedly have such experiences frequently and, to a certain extent, at will. Police departments on several continents have made frequent use of psychics in the solution of crimes. Archaeologists have employed psychics to assist in the location of buried sites and artifacts. Water dowsers have often been employed, sometimes surreptitiously, for the location of wells. Mining and oil companies have used clairvoyants to locate underground deposits. There has been active interest in military applications of remote viewing and other psychic phenomena, especially during the Cold War, on both sides of the Iron Curtain. In the United States, various military and intelligence agencies have sponsored, supported, or conducted research into the strategic applications of these capacities.

Yet as an area for scientific research, interest has been limited and financial support still more so. Among the attempts to set up repeatable experiments to demonstrate and study such phenomena, the best known in a historical sense are those initiated by J. B. Rhine at Duke University (continued at the Foundation for Research on the Nature of Man) in Durham, North Carolina. The Parapsychological Association, which has concentrated on such attempts, was admitted to the American Association for the Advancement of Science (AAAS) in 1969.

2. *Those where the state of a person's mind appears to exert a direct effect on the physical environment.* These include simple *psychokinesis* (in which the state of mind apparently results in something being moved or physically affected at a distance and without physical intervention of the ordinary sort—for example, some of the reported metal-bending demonstrations); *levitation* (of oneself); *teleportation* (apparent disappearance of an object at one location and simultaneous appearance at another); *materialization* and *dematerialization; thought photography* (in which a held image or mental state apparently results in an image on photographic film); *psychic healing* and *psychic surgery.*

To most scientists these reported phenomena comprise dubious claims that can largely be explained by "natural" causes. Some of the sponta-neous occurrences are well attested by reliable wit-nesses, and are less easily disposed of. (See, for example, cases reported in Brian Inglis, 1992 and Jule Eisenbud, 1967.) The most impressive laboratory data is that reported by Robert Jahn (Jahn and Brenda Dunne, 1987) at Princeton on mental influencing of an electronic random-event generator.

3. *Events or phenomena that seem to imply survival of physical death.* Classical examples include *mediumistic communication or channeling* (apparent communica-tion with discarnate beings, sometimes accompanied by physical or quasi-physical phenomena), *polter-geists,* and *apparitions.* While the word "channeling" is used these days quite loosely, and often with an apparent premium on gullibility, some of the classical cases of mediumship were quite impressive in their evidential quality. (See, for example, Rosalind Heywood, 1974; Brian Inglis, 1992.) Some of the materials that purportedly are authored by discarnate beings appear to be of extraordinarily high quality (such as the "Seth Materials" and "A Course in Miracles," discussed by Arthur Hastings, 1991). Furthermore, as will be detailed below, the subject of "survival" has not been laid to rest; additional evi-dence has appeared of a different sort.

Research in these three categories, which we have grouped together under the heading "meaningful coincidences," has been largely unappreciated by the mainstream scientific community. Since the phenomena are so contradictory to the scientifically known reality picture, they must be either "meaningless coincidences" or explainable by fraud and collusion. However, William James' dictum of "radical empiricism" would suggest that the problem lies not with the phenomena but with accepted scientific epistemology. As Jung suggested, the data and experiences of synchronicity appear to be strong enough as to imply some sort of "acausal connectedness" that demands investigation.

The issue of unembodied existence

The implications of the "survival" topic, especially as regards the widespread fear of death, are sufficiently significant to justify a slightly fuller report.

The medieval worldview was characterized by a continuum between this world and the next, such that the question of continuation didn't even come up. This continuum was shattered by the scientific revolution, so that by the mid-nineteenth century there was a near-total discrepancy between a religious worldview within which the survival issue was presumably resolved and a scientific worldview within which the question was groundless. Interest in the survival question peaked around the turn of the century and waned to a mere trickle after World War I. There was a slow resurgence of interest beginning in the 1960s, and a fresh look at the question seems to be emerging in the 1990s.

Much of the older evidence for the survival hypothesis centered around the phenomenon of mediumship, wherein a person in an altered state of consciousness appears to be able to receive communications from discarnate entities, and on occasion to evoke such physical manifestations as raps, table tipping, ouija board influencing, slate writing, and the like. Messages came in various ways. Some were oral utterances by the medium, taken down by a recorder. Others came in the form of automatic writing. A few were inscribed on closed hinged slates (of the type that used to be commonly used by schoolchildren) in which a slate pencil had been inserted, and the closed slates held by the researcher or placed under heavy objects to eliminate any possibility of fraud. (On careful examination, the particles of writing material appeared to have been deposited on the slate face, rather than rubbed off the slate pencil in the normal way. Of course, the idea that writing could take place without a writer to move the pencil was not accepted by skeptics, but there seem to have been adequate critical observers to give the reports some credibility.)

Eventually all of this activity attracted the serious interest of such scholars as Sir Oliver Lodge and Frederic W. H. Myers in England and William James in the United States, and led to disciplined investigation and the creation of professional societies, the most prestigious being the Society for Psychical Research (SPR), formed in 1882.

Myers's investigations in England were outstanding, and toward the end of his life in 1901 he summarized the evidence for survival in a landmark two-volume work entitled *Human Personality and Its Survival of Bodily Death.* He and his fellow researchers were consistently frustrated by the difficulties of studying mediumistic communication. Thus he half-jokingly promised his fellow workers that when he died he would devise an experiment that would leave people in no doubt as to his identity and survival. Beginning shortly after his death, and continuing for three decades, there were a remarkable series of communications purporting to come from him (with a few as well from his colleagues Edmund Gurney and Henry Sidgwick, who had also died by this time), which became known as the "cross-correspondences." These scripts came to a dozen mediums, living at various locations on three continents. They comprised fragments of messages, including fragments of classical quotations, which were clearly incomplete in themselves but when assembled at the SPR office in London fit together like pieces of a jigsaw puzzle (Inglis, 1992).

Myers's attempt to bring afterdeath experience into the reach of science did not, it seems, stop with his death, nor even with the cross-correspondences. Over twenty years after his death, a sensitive in the north of Ireland named Geraldine Cummins began to take down through automatic writing lengthy scripts attributed to the deceased Myers. These were published (with Cummins identified as author, but with a foreword explaining why she believed them to be transmissions from Myers) as two books, *The Road to Immortality* and *Beyond Human Personality* (Johnson, 1954). They contain a fascinating report of his afterdeath experience and his mapping of the afterdeath possibilities, the latter being broadly similar to mappings that have appeared before and since from other sources.

Basically, death appears less as an extinction than as awakening to "where one is all along." We don't go somewhere at death; we are already there. In general, the center of awareness shifts at death from the physical to the higher planes. Immediately after death individual experiences are as different, one person from another, as they are in earthly life. There may be a period of confusion and/or sleepy resting, or we may visit whatever was our idea of "heaven" while on the earth. When the soul is ready, learning resumes; the journey to greater awareness continues. Our consciousness is of the same quality as that which pervades the universe, and our awareness of that universal consciousness is potentially without limit. Our sojourn on Earth has been, in Wordsworth's words, "a sleep and a forgetting"—a temporary unawareness of our true nature. Learning does not stop with the death of the body; the path to higher awareness is never-ending.

All the work with mediums over many decades faced the obvious problem that, whatever the original source of the communication, there was no way of telling how much it had been corrupted by bubbling up through the medium's unconscious mind. This problem plagued all of the researchers from Frederic Myers on, and was a source of continuous frustration, even when there seemed to be something significantly evidential in the messages received.

As if in response to this problem, shortly after magnetic tape recorders became widely used, in the 1950s, messages began to appear on various tape recorders that purported to be from discarnate beings. This was "hard" evidence, presumably uncontaminated by the mind of some medium. Some of these messages were from persons who, prior to their death, were deeply involved with research on the survival issue. In still more recent times, as other technologies became available, these communications have extended to involve television screens, videotape recorders, and words and images scanned into computer disks; to include real-time two-way communication; and to include photograph-like images as well as verbal messages. All of this would seem on the face of it to constitute a totally preposterous claim, yet some of these communications, collected by researchers in at least six countries, comprise intriguing evidential significance (Mark Macy, 1995). They even suggest that further progress will be made through the active collaboration of researchers on *both* sides of the curtain we call death.

There is also clinical evidence seeming to support the survival hypothesis. Much of it comes from the new psychotherapy fields of "spirit releasement" (Baldwin, 1993) and "regression therapy" (Lucas, 1993). The former has to do with the freeing from a condition that was once called "spirit possession," now widely assumed to have been a prescientific notion without substance. In the latter, clients supposedly resolve present problems through understanding their origins in a past life. Some validity to this is offered by the painstaking work of Ian Stevenson in verifying the information in childhood memories of past lives (Stevenson, 1987).

Needless to say, although both of these kinds of therapy are fairly well established, neither the concept of recollection of past lives nor the possibility of spirit attachment is generally considered to merit scientific credence, since they imply things such as reincarnation and discarnate intelligences. In weighing this lack of official endorsement, however, one should bear in mind that the concept of the unconscious mind had become a widely accepted basis for psychoanalysis and other psychotherapies a full half century before it gained acceptance in strict scientific circles.

In discussing these two research areas (nonordinary states of consciousness and "meaningful coincidences"), there is no claim to have demonstrated conclusions. Our purpose was merely to indicate directions of exploration that some researchers, at least, are taking seriously.

CHAPTER
SIX

Social Implications

In the context of today's worsening global situation and our imperiled future, perhaps the most important feature of the described new outlook of science is its provision of a prescription for long term, high quality survival and a way out of our current global predicament.

—Roger Sperry (1994)

The time has arrived in which we have to realize that we are all parts of a single organism and develop some new kinds of responses and relationships.

—Jonas Salk (1983)

HARMAN:

We have both agreed that we would like to explore some of the ways in which changing biology would affect society. Certainly one of the most radical concepts we have been talking about is the finding of something like intelligence throughout the universe. I think I'd like to start by describing how I came to view these matters the way I do—because I believe that the essence of my story is not as uncommon as one might at first thought assume.

Like many people of my generation (I was born during World War I), I grew up with science being an unquestioned authority and the nature of reality. I studied science, and ended up teaching applied science and engineering at Stanford University. Then at age thirty-six I was tricked into going to a two-week retreat in the Northern California redwoods. I say "tricked" because if I had known what was likely to happen to me there I probably would never have gone. It had been represented to be a nonreligious discussion of ethics and life principles. I found it was a great deal more—more experiential, in particular. This was long before the era of sensitivity training and encounter groups, let alone *est* and its many cousins. But here we were, spending long periods meditating with soft classical music background, doing art with the left hand, and, especially, sharing with one another at length what we had felt deeply about, both positively and negatively, throughout our lives.

I found it all pleasurable and in some way fascinating. Toward the end I realized that I felt betrayed by the leader who had enticed me to enroll. The reason I felt this way was that he had revealed that he believed there is something to psychic phenomena and the power of prayer, and I felt that someone with his education (he was a professor of law at Stanford University) ought to know that science had disproved that hypothesis long ago.

The last day I burst into tears when I was attempting to describe what I felt I had gotten from the seminar. I was aware that the sobbing had more to do with joy than with sadness, but I couldn't for the life of me give a good explanation of what lay behind it. It was as though some part of me was signaling that it was about time I got my life off dead center.

Clearly there were aspects of life they never told me about in school. I set out to find out about them, without really knowing how. As the years and decades went on (I'm a slow learner), it gradually dawned on me how much of how we see and experience ourselves and the world around us depends upon the beliefs we have internalized along the way. Encountering the mysteries of hypnosis was a part of this. If, through accepting the suggestions of a hypnotist I can be brought to see things that aren't there, to fail to see things that are present to everyone else, to display strength I never had before, to be unable to lift the smallest weight, to create all the signs of a burn when touched by a cold object I had been told was hot, or to walk on burning coals without blistering my feet—if all of this, then how much more it must be true that I see the world the way I have been taught to see it, by that master hypnotist called "our culture." We are all culturally hypnotized from birth! It explained so much!

I learned how this inner programming can be changed through what amounts to autosuggestion. If I suggest to myself, in a state of deep acceptance, that the world is full of wonder, or that I already have what will make me happy, it could really change things. It made sense. Also, my own meditative experiences were going deeper, and the profound connectedness of everything was becoming something I really believed. But if everything is connected to everything, why should it not be the case that if I imagine something, prolongedly and/or repeatedly, as vividly as possible, and I repeat or hold that imagining, that something might be manifested? I found that this was indeed a claim that had been made repeatedly, whether the process were called "affirmation" or "prayer."

However, something about all that was troubling. Let's suppose it's true that if I choose to pray for something, this can influence that something coming into being. But *who is doing the choosing?* If I am intimately connected to the Whole, do I really want what I think I want, or is that just the ego having its way? Don't I really want exactly what the Universe wants, since ultimately that is who I am? The prayer, "Thy will be done" began to make more sense than anything else did.

That conviction was amplified by a "near-near-death experience" at around age 60. I wasn't physically near death, but I was in a profound depression. For the first time I understood what people go through with psychological depression. This episode seemed to have to do with the deep realization—I felt it somewhere in the region of my stomach—that my life was approaching its end. It really felt like a death experience, and then one morning I walked up the hill to see the sunrise and it all lifted. Life was joyous again. But something was clear (at a deep feeling level) that had not been before. All the things that I had been taught all of my life were of value clearly were of no value at all, for in a few instants they would be gone. But one thing is of value—and only one. Alan Watts had called it "The Supreme Identity"—the identification with the Divine.

With this realization the journey seemed complete. I couldn't believe that the secret of life is so simple! If it is, why are there so many complex treatises written about it? Why do we all make recognizing it so difficult?! All I can say here is that two decades later, it still seems simple.

And now another kind of prayer began to make sense. If I am really co-creator of the Universe, then the power of the prayer of gratitude is obvious. It's not just that by focusing on what is beautiful, and good, and true, and experiencing a deep feeling of profound gratitude, I change the way I *see* the world. I actually change the world! It may be the most effective kind of service I can render.

I am now (they say, statistically speaking) in the twilight of my life. At any rate, the great adventure of death cannot be too far off. That fact in no way diminishes the delight of simply being out in nature, or playing with my great grandchildren, or enjoying the quiet comfort of a marriage that has lasted nearly six decades.

I could never have been persuaded, in my early years, that growing old can be thoroughly delightful. But in truth it is, every moment of it. It's fun to realize that at some deep level in ourselves, we *choose* to age and die, and to watch the process with as much joy and fascination as a new father might delight in watching the birth of his child. It becomes more obvious each day that I am merely a part of the Whole, that the deepest pleasure in life is serving that Whole, and that subtle and manifold are the ways we can discover to that end.

Out of that experience, which I have taken the liberty of sharing quite intimately, I am brought to several conclusions. One is that real learning does not come solely through assimilating knowledge; it involves coming to hold one's conceptual frameworks sufficiently lightly to allow in experiences that don't fit well with the existing frameworks. More and more people are coming to this attitude regarding the prevailing science, despite its impressive successes.

Another conclusion is that if our belief systems fundamentally change, through whatever process or experiences, our perceptions and everything else about our lives will change. That will be true individually or collectively.

Which is to say, the consequences of adopting such a holistic concept of biology as we have been talking about are hard to anticipate, but they will be profound.

SAHTOURIS:

Your story is wonderful, Willis. Thank you for telling it. I'm not so far behind you on this fascinating journey and I agree wholeheartedly with your urging that we "hold our conceptual frameworks sufficiently lightly to allow in experiences that don't fit well with the existing frameworks." To me, formulating, expanding, and revising my worldview is the most exciting thing about being alive. I was thrilled to learn the scientific worldview while still relatively young, and I preached it to anyone who would listen; later I suffered its constrictions like a suit too tight when one expands, and over the years I have tried out many new ideas, many new ways to make it coherent. I intend to go on doing that all my life, welcoming ideas better than my current ones, willing always to be proved wrong.

The biggest shift I've had to make is this shift from seeing consciousness as an emergent phenomenon to recognizing it as being there all along. It wasn't easy; yet it is so obvious once you make the shift. I couldn't have done it with only the science and spiritualities of my own culture to draw on. I needed to see how many other intelligent and questing human minds throughout history and its many cultures had reached the same conclusions. I needed to embed myself in indigenous cultures that lived it. I really thought Western scientists were more advanced than all of them; I had to learn to laugh at my own—our own—arrogance and accept that we were the backward ones. It's not so hard when you gain a far more exciting and wondrously magical world as a reward.

If we come to see the universe, and, more locally, our planet with its ecosystems as intelligent, it seems inevitable that our esteem for nature will rise. We will no longer see other species, and life itself, as the consequence of accident. That alone could be very beneficial, because elevating our respect for nature while adopting a little more humility ourselves might save us as a species.

Anyone looking at us from the Moon will see only one indication of our human presence on the Earth: the rapidly growing deserts we have been steadily creating since our invention of large-scale nomadism and agriculture, much intensified now by our industrial age. Perhaps pollution begins to show in our cloud-cover as well. Radio telescopes from space would also show man-made electromagnetic activity so great that Earth might be taken for the twin star of our Sun.

We modern humans, in designing and building our economic and political "machinery," considered ourselves separate from the rest of nature and even in control of it, but, in truth, our social organization is as biological a phenomenon as that of any other species and it is high time we understood that.

A biology that sees all nature as composed of co-evolving holons in holarchies will quickly reveal much about the nature of humanity itself as one such holon—or as a holarchy of individuals, families, organizations, communities, nations, and other holons. Through this understanding of ourselves, we will gain profound insights on where we succeed and where we fail as living systems embedded within other living systems. As I gained this contextual perspective myself, it often brought to light instructive parallels between our evolution as a human species and the evolution of cells and multicelled creatures.

For example, if we look back once more at the evolution of eukaryotes from tentative cooperative unions or communities of prokaryotes with diverse lifestyles, we can see that their intracellular biochemical or physiological production, transportation, and communication systems were invented prior to their communal phase, as were some intercellular systems. Margulis and Sagan (1995), for example, describe the proton motors in the membranes of spirochete-like bacteria as being complete with rings, bearings, and rotors that spin at fifteen thousand rpm, driving attached protein filaments (flagellae). I mentioned earlier how they trade DNA information as frenetically as stock traders on the floor. Some could move rapidly, some could manufacture food, and all of them used the same ATP energy currency. In other words, as in the human world today, the technology that makes cooperative living possible preceded their cooperation and their union into nucleated cells.

Margulis and Sagan, speaking of bacterial evolution, say, "Evolution is no linear family tree but change in the single multidimensional being that has grown now to cover the entire surface of the Earth. This planet-sized being, sensitive from the beginning, has become more expansive and self-reflexive as, for the past three thousand million years, it has evolved away from thermodynamic equilibrium." (p. 73) Margulis calls us humans, along with fungi, plants, and other animals—all made of the bacterial cooperatives we call eukaryotic cells—"powerful prodigies of bacterial life." And indeed, now, it is our turn, as humans, to form ourselves into a cooperative planetary "being."

In just the past few hundred years of our whole evolution as tool-making humans, we created production, transportation, communications, and currency technologies that have changed the whole planet and that are uniting us into a new kind of planetary holon. Without strife, we have built an efficient worldwide system of mail, telephone, and electronic communications; a worldwide air, sea, and land transport system; a global money exchange, and a rudimentary, tentative world government, in the form of the United Nations. Our multinational corporations are producing and distributing their products globally and we are continually working on international agreements of all sorts. Yet we have hardly even been aware that we were evolving into this holon—this global body of humanity that is evolving as naturally as did our physical bodies and the cells of which they are composed.

The lack of biological awareness of ourselves as one species holon within a planetary holarchy made it possible for us to be so focused on our intraspecies conflicts that we created nuclear weapons without seeing them as a means of committing species suicide. Nor were we intentionally destroying and polluting the very environment from which we draw our nourishment when we created our industrial lifestyle. We are just learning that these consequences of our industrial activity are actually threatening our survival. The new biology can help us be far more conscious of this process and far more realistic about the dangers posed to all humanity by its intraspecies conflicts and economic practices.

Many people are afraid that ecological practices will put our economy into jeopardy—that jobs will be sacrificed for the sake of endangered species. It is natural in our culture to think in this way: that for something or someone to gain, something or someone else must lose. We are accustomed to the win/lose economics of our globalized industrial culture and do not yet understand what Hazel Henderson has long called "win/win economics," the kind demonstrated by nature, which we ourselves must adopt if we are to survive in health. Sustainability, in its essence, is about this necessary shift to a win/win economy that would benefit all humanity as well as the other species on which human life depends.

As I said earlier, we must bring economy in its original Greek meaning of "household law" together with ecology as "household organization" (see Glossary, p. 129) instead of pitting them against each other as some sort of adversaries. Consider the biological processes of our physical bodies: Their ecology is their organization—the interrelation of the skeletomuscular system, digestive and perceptual systems and so on. Their economy of food intake, of cellular, endocrine, plasma, etc. production, of materials and product distribution and the recycling and elimination of wastes, is ruled, or regulated, by the guiding nervous system, via the endocrine and blood systems. As long as the body is healthy, there is no conflict between its ecology and its economy. It is practicing a win/win economy in which all parts contribute what they have to offer and all parts benefit equally from the collective economy. No part of a healthy body gains its health at the expense of other parts; there are no such things as rich and poor organs or preferred parts.

In our new biology terms, our bodies and the eukaryotic cells they are composed of are like mature ecosystems whose internal competitive negotiations have established mutual consistency among all the holons of their holarchies—in common parlance, the parts of wholes.

Our contemporary human society is an interesting test of the Darwinian evolutionary model that has guided its economic organization. We have assumed that competitive individualism, with profits as a bottom line, in leading to a healthy "survival of the fittest" would somehow benefit us all. But this model leads to a ruthless elimination of all but the most aggressive competitors and those who can eke out their existence in noncompetitive roles or in support of the fittest. We are now reaping the unfortunate effects of this model as megacorporations flourish at the expense of a labor force "downsized" or replaced by competitively cheaper labor in other parts of the planet.

The new biology would show us that mature ecosystems do not evolve toward the sole survival of the most aggressive and clever, as they sometimes do in early stages of self-organization or when invaded by very aggressive "outsider" species. Rather, they evolve toward intelligent and cooperative mutual support communities in which every species has a valid and valued role. In such a community, the bottom line is not profits, but output useful to other species in a virtually 100 percent recycling economy.

It would be wise for us to reconsider the whole notion of pollution and cleanup. Nature is fundamentally and necessarily based on recycling. Because the need to recycle our human products, lest they choke us out of existence, has become so urgent, a new branch of biological science is finally looking at nature's recyclers. It has now been estimated that sixty percent of all species are "recyclers." While this new science at last vindicates the vultures, worms, and microbes we have looked down on for so long, it is actually misleading. The natural world is not divided into producers and recyclers; all species are both.

In a mature, balanced ecosystem, as I just said, there is no waste, no pollution, no cleanup. The principle of mutual consistency suggests that a healthy species ensures its survival by putting out only quality material. "Quality material" is something useful to others. It is our industrial culture, immature from an ecological/evolutionary perspective, that thinks in terms of polluting wastes and cleanup. But it is becoming increasingly evident that adding more technology to clean up ever increasing wastes is a losing battle and cannot lead us to sustainability. Rather, as Paul Hawken urges (1993), we must go back to the drawing boards and redesign all our products so that they are either consumable or recyclable. It is not a matter of saving the environment, he says, but of saving business. Hawken proposes that if companies producing nonconsumables were only allowed to lease them, and not to sell them, thus ultimately being responsible for their disposal at great expense, they would quickly redesign them to be recyclable. Karl-Henrik Robèrt's Natural Step program, originating in Sweden and now spreading to other countries, including the United States, is a further development of these ideas.

Humans are not the first species to threaten their own and others' extinction by way of resource depletion and pollution. Ancient bacteria forced each other to survive similar crises by reorganizing themselves and their living systems repeatedly. Species living now can exist only because the Earth spent billions of years burying atmospheric carbon in forests and underground. Cutting and burning these forests and fossil fuels reverses the planet's system for keeping atmospheric conditions and climate conducive to the health of its species.

Our current way of life is now recognized as nonsustainable in the long term. It is the way of an immature species that gobbles up all available resources, like the wild plants that take over land along our highways or in abandoned fields, where we have destroyed mature ecosystems. Technological production is natural to the human species, but must be reevaluated and revised in a goal-setting context of healthy survival. Mature ecosystems can clean up considerable human pollution, if they remain healthy and are enlisted in cooperative ways. Plants, for example, can clean up water polluted in a staggeringly wide number of ways, even oil spills, heavy metals and nuclear wastes (Sahtouris, 1991). But the destruction of forests, seashores, water tables, arable land, oceans, the ozone layer, etc., make it nearly impossible for the Earth to perform that cleanup.

We have gained considerable understanding of living systems and their dynamic ecological balance; it is now up to us to work with life for life, eliminating waste as a concept and as a reality. Eliminating waste is more generally about reducing our impact on the planet and giving up the wasteful consumer lifestyle in which we define ourselves by totally unnecessary accumulations of goods. It is also about implementing accountability for restorative behavior and profitably using renewable or permanent energy sources to make what we do need, as Amory Lovins of the Rocky Mountain Institute was one of the first to demonstrate can be done. In Chile, a study showed that more energy could be saved through energy efficiency measures than would be produced by six new dams planned for the Bio Bio River, but there is virtually no chance of stopping lucrative, if destructive, big dam-building processes all over the world until we see the entire question of sustainability in the broader perspective of a new planet biology.

If we accept the notion of the living Earth, and the body of humanity as an integral part of it, then we have no choice but to implement a healthy human win/win world, which is to say, a sustainable world, a balanced economy of equal partners, as Al Gore (1992) among others suggests, rather than an economy in which some nations or corporations gain at the expense of others. Gore recognizes that "any such effort will also require the wealthy nations to make a transition themselves that will be in some ways more wrenching than that of the Third World, simply because powerful established patterns will be disrupted . . . the developed nations must be willing to lead by example; otherwise, the Third World is not likely to consider making the required changes—even in return for substantial assistance."

All this ferment reflects our growing understanding of living systems. And in this light it is interesting to consider historian Arnold Toynbee's observation, after studying twenty-one collapsed civilizations, that what they had in common was inflexibility under stress—refusal to change the percepts on which they were founded—and the concentration of wealth into few hands. Our global civilization now stands on the brink of either self-destruction or self-organization toward healthy and harmonious community. I cannot envision anything that would better assist our going in the latter direction than the new biology we are proposing.

HARMAN:

Most important, I believe, is the "new story" that this new biology points to. The materialist picture of our accidental origin and random, meaningless evolution contributed to modern society's confusion about values and ultimate purposes.

The most fundamental political question of our time, it seems to me, is what picture of reality shall guide our lives and our society, and who shall control that picture.

SAHTOURIS:

I'm in complete agreement, and of course, whoever gains that control I would hope will be guided by our parent planet's four to five billion years of experience in the organization of viable living systems. Let me refer back to those principles of organization that I listed at the end of Chapter Four and also quote from my discussion of ecological ethics (1996) to show what a shift from Darwinian to holarchic evolutionary biology would mean for society at large. Then I'd like to compare a quote from David Korten's article, "Economies of Meaning" (1996), to show how very similar a guiding proposal is emerging within the business community you know so well. I believe this will show clearly that the new biology is impacting other sectors of society before it is even established in science!

From my own discussion:

. . . The competitive exploitation of resources and labor by the [entrepreneurs] as they built an industrial world was thus justified on the [Darwinian] grounds that it was natural . . . a common human ethics is what we now need more than anything else—an ethics to guide us in our behavior toward one another and toward the natural world to which we belong. Our basis for such an ethics is very different now that we no longer see nature as just a bloody battleground for competitive struggles over limited resources. Competition is merely one aspect of nature's creative organization into mutually consistent holons within holarchies. What we see clearly is that every holon's health depends on the health of the larger holons in which it is embedded. Thus every holon, in looking out for itself, must also cooperate with other holons to help look out for their larger holon's interests.

This, as we said, is the heart of ecological ethics—the self-interest of any particular holon, whether a cell, a body, a society, a species, an ecosystem, or a whole living planet, balanced in mutual consistency of the whole and all its parts. For us this means recognizing how much we affect the living planet of which we are part and on which our continued existence depends. To look out for our own interests requires that we know the interests of our whole environment, which means our whole living planet. Our free choices, in order to serve our own long-range interests, must serve those of other species as well, for natural ethical behavior is that which contributes to the health of the whole (Earth) system.

Our history has brought us to shortsighted . . . warfare, hatred, distrust, and reckless destruction of our own environment. We have long-standing habits of believing that all nature is human property, and so we take land and resources from one another for reasons of profit. It is high time for us to realize that maximizing individual profits minimizes human social stability and welfare, while maximizing common profits destroys our natural life-support system. If we want to survive as a species we must learn to change our ideas and our lifestyles to live in a balanced recycling economy like the rest of nature.

In fact, it is high time to realize that all our old habits and vested interests, even if they form our individual and national identity, must be fundamentally changed. The changes required are deeper and more far-reaching than any revolutionary leader has ever demanded or even dreamed of demanding. And yet we can make those changes peacefully, and everyone can win.

The problem is that they cannot be made at the point of a gun. They must be made voluntarily, and that is perhaps more difficult. The profit motive is so ingrained in Western society, for example, that scientists have actually criticized nature on the grounds of unprofitable inefficiency, pointing out that photosynthesizing plants use only a small fraction of the energy available in sunlight. Can such people learn to appreciate the fact that plants extract exactly as much energy as they need for themselves and to keep their environment's careful balance of energy exchange?

...If we agree to consider ethical human behavior as whatever we sincerely believe, on our best knowledge, to be healthy for ourselves, our families, our species and healthy or at least harmless for other species, for the environment, and for our planet, then we have such a guide.

Now let us compare Korten (1996) as he summarizes the moral premises of our corporate industrial society:

- People are by nature motivated primarily by greed.
- The drive to acquire material wealth is the highest expression of what it means to be human.
- The relentless pursuit of greed and acquisition leads to socially optimal outcomes.
- It is in the best interest of human societies to encourage, honor, and reward these values.

...this is the essence of the value assumptions underlying most contemporary market theory. Unfortunately, economic policies driven by these deeply flawed moral premises create a self-fulfilling prophecy by rewarding dysfunctional behaviors deeply detrimental to the healthy function of human societies, as we now see demonstrated all around us.

Our development models—and their underlying myths and values—are artifacts of the ideas and institutions of the industrial era. The corporation and modern state have been cornerstones of that era, concentrating massive economic resources in a small number of centrally controlled institutions. They have brought the full power of capital-intensive technologies to bear in exploiting the world's natural and human resources so that a small minority of the world's people could consume far more than their rightful share of the world's real wealth.

Economic globalization has served to advance this exploitation of the earth's social and environmental systems beyond their limits of tolerance, by freeing errant corporations from restraints to their growth, their ability to monopolize ever larger markets, and the use of their economic power to win political concessions that allow them to pass on to the community ever more of the costs of their production. It has delinked corporations and financial markets from accountability to any public jurisdiction or interest, contributed to a massive concentration of financial power, and richly rewarded those who place the values of acquisition, competition, and self-interest ahead of values of simplicity, cooperation, and sharing.

We are not limited to choosing between markets or governments as the instruments of our exploitation. Nor is there need to eliminate markets, trade, private ownership, the state, or even the institution of the corporation. Rather, it is a matter of creating a new architecture for each of these institutions appropriate to the values we believe a good society should embody and nurture. This creative task belongs neither to corporations nor to states, which are incapable of questioning the assumptions on which the legitimacy of their present institutional form is based. It belongs to citizens—to the people whose interest and values the new architecture is intended to serve. It is people rather than corporations or other big-money interests that appropriately set the terms of the economic and political agenda.

Citizen groups throughout the world are already actively engaged in the experimental creation of economies of meaning aligned with life-affirming values. Powerful formative ideas are emerging from these initiatives. For example, millions of people in the voluntary simplicity movement are discovering the good living is more fulfilling than endless accumulation and consumption. In a healthy society, a life of material sufficiency and social, cultural, intellectual, and spiritual abundance can readily be sustained in balance with the environment.

Others are learning that there are alternatives to a global economy that inherently fosters inequality and global competition among local people and communities. They are demonstrating such possibilities by building strong and self-reliant local economies that root resource management and ownership in democratically governed communities and recognize that all people have an inherent right of access to a basic means of creating a livelihood. Such economies are an essential foundation of healthy societies able to engage in cooperative and caring exchanges with their neighbors.

These are lessons with profound implications for a politics of meaning. In large measure, societies express and sustain their cultural values through their choice of economic structures. The fact that our present economic system value and rewards greed, gluttony, and disregard for the needs of others didn't just happen. It is a consequence of conscious acts of choice—poorly informed though they may be. It is equally within our means to create a globalized system of localized economies that thrive on life-affirming values of sufficiency, caring, cooperation, and reverence for life. It is a matter of adequately informed collective political choice.

It is of interest to note that not a few nonindustrial cultures have long held values and purposes consistent with deep understanding of living systems. Many of them were sustainable for thousands of years until aggressive empire builders, more recently known as colonialists, destroyed them. In nature this can happen with the sudden introduction of a predaceous foreign species, such as the eucalyptus tree or the fungi that wiped out the American chestnut and elm trees. It may also be seen as similar to the ancient bacterial phase just prior to the formation of nucleated cells.

At that time, Margulis tells us, relatively fast-moving, high-energy respiring bacteria, which had evolved in response to the massive production of oxygen by photosynthesizing bacteria, had run out of free food in the form of natural sugars and acids. Under duress, they invaded the larger, more sluggish fermenting bacteria and "ate" them from the inside as their numbers multiplied within the colonized "hosts," stretching their cell membranes to enormous proportions.

This "bacterial imperialism," as I have called it, must often have come to a dead end when the colonial resources were used up. But it also set the stage for bacteria to live within each other's bodies, which made possible the intelligent shift to a cooperative division of labor. For example, the respiring spirochete-like bacteria could attach to the outside of the communal host and push it into sunlit waters where photosynthesizers, taken aboard, so to speak, could manufacture food for the whole enterprise. The spirochete-like members of this community evolved into cilia; the photosynthesizers into chloroplasts; and so on—with each member contributing some of its DNA, as we have said, to the evolving nucleus where all could draw upon it.

It is an interesting aside here that only a few years ago a Nobel Prize was given for the discovery of "gibberish" or "junk" DNA—labels no longer used as microbiologists discover the functions of what was previously mysterious. Nature wastes nothing, as far as I can see. The term "junk" is inappropriate, except when humans produce nonrecyclable materials. As yet I do not believe we fully understand DNA. If indeed it turns out to do no more than code for amino acid and protein sequences, then it can no more account for the organization of a body than can a shop full of yarn account for the organization of a sweater. At the cellular level, Bruce Lipton (1995) sees the cell membrane as having a far greater role in making decisions and carrying out cell functions, including the way in which DNA is used, than the DNA itself. He also points out that ancient eukaryotes (sometimes still called protozoans, though Margulis insists there are no single-celled plants or animals) evolved all the physiological systems present in humans: digestive, respiratory, excretory, integumentary, reproductive, cardiovascular, musculo-skeletal, even immune and nervous systems, and, like Margulis, he sees them as aware or conscious. This is very relevant in our new biology, which forces us to look to all levels of a holon and its holarchies at once.

Recall my description of how nature works out the balance between self-interest and interest beyond self, as for example in our own body cells, each of which is a holon in its own right as well as part of larger holons in the body holarchy. Each cell contains the genetic complement of the entire body, and thus has access to it if changes need to be made. And each nuclear genetic resource library may itself prove to be organized as a holarchy of holons representing interests at all levels from the whole body to those of each individual cell. This is speculative, yet we do know that our whole bodies are cloned from a single cell and that each cell "switches on" the genes that concern its particular organization and work. In some sense, information relevant to the entire holarchy, from cell to organ, organ system and body levels of organization, must be present in that first fertile cell, perhaps as a high-frequency field that is an aspect of the membrane, the DNA and other parts together. Does this field expand as new successors to this cell are cloned? Does each new cell carry its own field? Or do both occur simultaneously, as seems to make most sense—the holons and holarchies both having field aspects.

We know that there is communication and trade among neighboring body cells and that materials and information are exchanged between individual cells and the most widespread parts of the body. This entire system unfolded during embryonic development in such a way that each level of the physiological holarchy from cell to body looks out for its interests, and thus they are pushed or pulled into the cooperation I call "mutual consistency." Let us look at the social relevance of all this. As a native Meshika elder named Xilonen Garcia once said to me: "Anyone who knows how to run a household knows how to run a world." I think we could rephrase this as: Anyone who understands the principles of living systems can apply them to holons of all sizes, including human families, nations, or the whole human world.

If every cell in an organ worked for its self-interest, but the organ as a holon did not, the cells might kill one another off in competition. Surely they would be disorganized to the point where there was no functional organ. In the same sense, a society in which people looked out only for their individual interests, because they were not asked to, and did not volunteer to, do anything in the interest of their collective society, could not possibly be a democratic society. Democratic governments are set up to manage the public interest, to create public works and institutions, to limit free enterprise and to tax some of its profits to meet society's needs. But if the people do not hold the government accountable to the whole society, particular interests, such as large corporations that put profits ahead of people, will become overinfluential, as we now see.

Consider the opposite situation—where the organ or the society is so powerful a holon that it can demand the complete self-sacrifice of its cells or people in serving its interests. The cells or people thus enslaved would no longer be individuals in their own right. Science fiction writers have tried to imagine humans becoming robotlike parts of a dictatorial mechanical society, but real people resist becoming "cogs in wheels," and they stop functioning well. This is why communist countries have either failed, as in the Soviet Union, or discovered they must give people some opportunity to work for their individual interests if their societies are to continue, as in the case of China.

The point is that neither the individualism of capitalism nor the collectivism of communism will suffice as the sole basis for a society. Rather, we must take our clues from nature and organize our societies according to the principles of living systems, with continual efforts toward mutual consistency of holons in holarchy. (Margaret Wheatley, 1996, makes this point eloquently.)

The body of humanity has not yet evolved the truly impartial and cooperative world government it needs to coordinate its interests as a whole. Looking back at evolution again, we recognize that there must have been a number of steps in the transition from prokaryotes to eukaryotes as competition among individuals gave way to their cooperation as members of a new whole. We know that two of the most important steps were the formation of the nucleus from the DNA of the various bacteria living within the same cell walls—the nucleus that could organize the information needed to carry out the activities of the whole, and the formation of the more complex cell wall, with its active, intelligent regulation of everything entering and leaving the cell. The same step was accomplished when nervous systems formed in multicelled animals that had evolved from protist colonies in which different member cells did different jobs.

Something of this ilk is clearly happening as the body of humanity struggles to form its new identity. Since the close of World War I, nations have recognized the need for some kind of organization to process complex information and to coordinate and balance national and international interests. The League of Nations was born, then the United Nations. Although the U.N. accomplished much in the way of promoting peace and developing programs and services, the competitive interests of member nations still dominate on important issues, limiting its powers and often preventing the smooth functioning of the U.N. itself.

The rise of official U.N. NGOs—nongovernment organizations, many of them grass-roots based—is an interesting development. It remains to be seen whether they will be incorporated into the present U.N. structure as it is reformed, or whether they will self-organize as a kind of parallel U.N., and history will work out which will become the main organization. A world government, if it follows the pattern of biological evolution, will not be autocratic or authoritarian but will become a world government in service to the needs and welfare of the body of humanity, like our individual nervous systems. If our human civilization is to survive, we have no choice but to solve this problem before long, completing our evolution into a worldwide body of humanity with a functional coordination system. Other features of our ongoing globalization are air travel and transport, the postal system, telecommunications (including the Internet), the Hague (international law), and various global agreements.

HARMAN:

I couldn't agree more. Just to add another voice, let me quote briefly from Hazel Henderson's *Building a Win-Win World* (1996):

Today, the most creative, energetic forces addressing the planetary problems of poverty, social inequity, pollution, resource depletion, violence, and war are grassroots citizen movements. Grassroots globalism is . . . now emerging as a third, independent sector in world affairs—challenging the domination of global agendas by nation states and transnational corporations. The global civil society, newly interlinked on the Internet and by millions of newsletters, is increasingly driving agendas of nations and corporations....

The rise of civil organizations is one of the most striking phenomena of the twentieth century . . . pressure on member-states [of the United Nations], from NGOs in both the North and South, resulted in a series of ad hoc conferences . . . [which] have amounted to a twenty-five-year effort to steer the course of economic development toward new values: ecological sustainability, poverty reduction, and recognition of the key role women play as the world's primary food producers, educators of children, and protectors of the environment.

SAHTOURIS:

That is a very important development indeed, and gives us hope that neither unwieldy obsolete governments nor greed-driven corporations can rule the people of the world if they awaken to their own power to change things.

Let's look at some historic examples of sustainable social systems that functioned by the principles of healthy holons, and see what became of them. Pre-industrial-age communities were usually small and well integrated into their ecosystems or bioregions. Many of them were able to function in good ecological and social balance for centuries and even millennia. They functioned much like cells and bodies with divisions of labor, their diverse parts contributing to each other's welfare.

Sarah James, in Rio de Janeiro for the Earth Summit in 1992, described her Gwich'in Indian culture in the northernmost inhabited village of Alaska as it was before contact with the white man. Her people's relationship with the caribou was sacred, and they were endlessly grateful for this wonderful animal that gave them everything they needed and wanted: food, bone and skin houses, boats, snowshoes, utensils, tools, clothing, drums, flutes, and sacred ritual objects. Their lives were rich—rich with family and community, warm homes and clothing, plentiful food, much time for ceremony, music, dance, storytelling and laughter, much reason for celebration and thanksgiving for their bounty. But when the white man came to them, he saw people living in forty degrees below-zero weather with only caribou to provide for their "meager" sustenance. He called them poor "savages." Sarah says with passion as she beats her caribou skin drum, "Well, then let's keep Alaska *savage!*"

Sarah was making a clear statement of preference for her traditional life of simplicity over the modern world that brought her people real poverty, along with the terrible dependencies of debt, alcoholism, and the glue-sniffing that has destroyed her own son's brain. She was also making the point that wealth is a matter of perception and priorities.

Helena Norberg Hodge (1991) has documented some of the last healthy communities of this type in Ladakh, the barren, high altitude "Little Tibet" of Kashmir, one of the harshest regions of Earth, where they existed intact up to the late 1970s. She shows how such cultures were systematically dismantled to bring them into the world economy.

The prosperous sustainable peasant communities of Ladakh had three-story white-painted houses, beautiful monasteries, irrigated wheat fields and gardens, herds of animals, festivals displaying their music, theater arts, brocades and silver, crops adequate to support their people in good health, and no poverty. Buddhists and Muslims lived peacefully together in these communities, with their deep spirituality and strong values. Yet, despite the considerable property and well-being she described, the barter economies of these communities counted as nothing in the national GNP. Only when the barter economies were undermined by the influx of the modern commercial world and men left these communities to work for a pittance in cities did the GNP go up. Ironically, the people were not better off.

How did they talk the men into leaving the spiritual beauty, communal harmony, and physical bounty of their villages for polluted, congested urban living? As in other parts of the world, roads were built and people were encouraged to stop producing their own food and goods by inundating them with subsidized grain and other cheap imported goods. Motorcycles, TV, and videos filled with guns, girls, and images of no-work affluence in the "modern world" came in to erode the economy and the values. People were told they were backward, that modernization would bring great benefits. Because of the influx of the initially subsidized grain, fields were abandoned; schoolchildren were systematically taught the values of a market economy and the importance of industrial development. In Ladakh, the downward social spiral happened in a single generation.

In Europe, Africa, Asia, Australia, and the Americas the process dates back further but was essentially the same. Most people of these formerly healthy living communities have ended up poor or destitute on barren land or in urban slums. They have become part of a world economy in which they serve as cheap labor and market outlets if they are lucky. Increasingly, they are left out of even these slim benefits, desperately poor in huge urban slums, on the edge of starvation throughout their lives, many never reaching adulthood. According to world futurist Rashmi Mayur and several TV documentaries, many millions of children in Bangladesh and India under the age of 10 are enslaved up to 19 hours a day, 7 days a week in factories making goods for export to the United States.

A 1994 *Atlantic Monthly* cover story by Robert Kaplan documents the devastating reality of this process for the entire world. Kaplan points out that to believe things are still well in the world one must ignore three-fourths of it. If we see this situation realistically, we know it is so unsustainable that it is leading us toward possible extinction and certain misery. We are, in fact, in the same desperate situation that the ancient bacterial colonialism led to, yet we can be inspired by their solution of nucleated cell cooperatives, which have survived and flourished now some two billion years in myriad evolved forms. They are so sustainable that no other kind of cell ever replaced them. The same cooperative solutions they found, without benefit of brains, are open to us now.

A sustainable world must be based on an understanding of what sustainability is. As nation-states become increasingly disempowered, and most of our transnational corporations become increasingly inhuman, it would be sensible for us to see ourselves again as living systems within living bioregions that have natural boundaries, such as watersheds.

In a bioregionally organized world, the various forms of scientifically integrated permaculture derived from indigenous and traditional agriculture could be used along with appropriate technology for other aspects of life, from communications to housing, medical care, etc. Local production would meet as many needs as possible for food and other goods, with imports determined by democratic discussions. Community would naturally become vital again in such settings, and local culture would flourish, while also exchanged with other regions.

Urban areas will still be desirable and necessary for efficient technological production and other activities and institutions, such as research institutes, whose knowledge could then be made available electronically to all. Many people are working on sustainable urban designs that integrate gardens and use clean, efficient energy and public transport. Curitiba, Brazil, a city of over one and a half million inhabitants, was recently featured by the *Scientific American* for its successful "design with nature" (Rabinovitch, 1996), implemented by its far-seeing former mayor Jaime Lerner when the city was much smaller. Curitiba is partly built on drained, reclaimed floodland that was restored by sound ecological design such that it also functions as flood control. The city planned for its own growth, which has been rapid but not overwhelming, as in other cities its size. Curitiba has a sensible, relatively low-tech public transport system that has successfully eliminated gridlock; many bicycle paths and parks; adequate recreational and leisure facilities; much low-income housing development; a high green area to people ratio; payments to households for trash; recycling; programs for children, including resources for children in danger and part-time jobs for youth, a unique system of incentives for positive citizen behavior; and a free University for the Environment. The *Scientific American* reported: "These innovations, which rely on public participation and labor-intensive approaches rather than on mechanization and massive capital investment, have reduced the cost and increased the effectiveness of the city's solid-waste management system. They have also conserved resources, beautified the city and provided employment." While Curitiba's problems are not all solved, it has made some of the world's most significant strides toward urban sustainability.

As ancient bacteria evolved into protists, there must have been far more failures than successes, instances in which unceasing exploitation and hostilities among bacteria multiplying within a single cell wall led to the destruction of the whole enterprise. Perhaps Curitiba is the first equivalent of a successful nucleated cell. But the body of humanity requires organization on the far greater global scale, and we humans cannot afford failure at this level, for we have only one chance. The common cell wall that binds us together is the boundary of our planet itself. If we understand the new biology, and the evolutionary pressure on us now to complete the organization of this new body, we can work at the task consciously and rapidly.

Some of my best insights on this score have come from outside the Western scientific paradigm, from various indigenous scientists and indigenous people in other walks of life. There is invaluable perspective to be gained from indigenous people. My Tewa friend Dr. Greg Cahete of the Santa Clara Pueblo in New

Mexico, author of *Look to the Mountain: an Ecology of Indigenous Education* (1994), points out that the white man isolates parts of nature in laboratories to study them because his purpose is to control them, while the native scientist goes out into nature to study it because his purpose is to integrate harmoniously with it. Only out in nature, where the phenomena under study are not torn from their contexts, is it possible to understand the vital interrelationships of parts or aspects of nature. Our scientists now understand the participatory nature of the universe intellectually, but indigenous scientists consciously practice their participation in a multilevel universe without compromising their ability to do good science. Western science has been very successful at developing technology, albeit often in inappropriate ways because it has not been a good science for our biological survival. The best science for that purpose is indigenous science, with its understanding of consciousness as inherent in all nature, and with its ecological knowledge, practice, and wisdom. With cooperation between indigenous and industrial cultures, we could develop a broader science to benefit all humankind.

HARMAN:

I believe that's correct. It's most interesting that recognizing the need for a new epistemology leads to some of the same places as does respect for the knowledge of the indigenous peoples.

SAHTOURIS:

Where then is our hope? The best clue I know is in the most powerful symbol of our age, one that has reached the hearts and minds of every human who has seen it: the singular and exquisitely beautiful image of our living planet Earth. It is clearly not a mechanical assembly of parts; from space its geology and biology are clearly inseparable, and if we could speed up its history to our human timescale, we would see clearly that it is alive, and by our arguments for the new biology of all the universe, including our planet, it is conscious as well.

This new biology we have discussed is an organic biology that accepts a multilevel reality in which consciousness and intelligence are not only legitimate areas of investigation but fundamental aspects of its epistemological and ontological framework. We have seen this biology's reflections of ancient Eastern and indigenous sciences, and shown how it could heal the historic rift between science and spirituality through the recognition of all nature and cosmos as alive and sacred, with profound implications for bringing human systems into harmony with the greater whole in which they are embedded.

HARMAN:

It's a sobering observation that the only societies we know to have learned to be both prosperous and sustainable in the long term are ones we used to label "primitive."

SAHTOURIS:

It certainly is. In fact, it is instructive to look to modern agriculture as an example of human practice that conflicts with principles of living systems, and to compare it with some of the ways agriculture was practiced by indigenous peoples. Our fossil fuel-dependent, high-tech agriculture is in crisis because of our failure to understand the nature of sustainable living systems, as the World Bank and other investors now begin to understand. In many parts of the world, genetically engineered and monocultured crops, intended to solve world hunger, have created disasters of soil and water pollution and erosion, soil death, and human death and illness due to chemicals. While great profits are made from them and the affluent world can eat whatever they want from anywhere in the world year-round, arable land is being destroyed and eroded at a frightening pace by unsustainable practices, and nonaffluent hungry populations are growing rapidly.

Biodiversity is essential in all living systems. Monoculture is as destructive and dangerous in agriculture as it is in human social systems. Physicist Vandana Shiva (1988) has documented the Green Revolution in India, tracing the development of nitrate-dependent agriculture to the need for maintaining the production and profits of nitrate explosives factories after the Second World War. Nitrate-dependent crops were deliberately bred for the much-touted Green Revolution. The statistics of this high-tech agriculture then accurately showed a greater yield of rice per hectare than traditional methods, but the measures were misleading because they ignored the fact that the same hectares were formerly producing not only rice but fish, pigs, vegetables, fruit, fertilizer, and mulch, without chemical input, on soil and in water that not only remained healthy but improved over time. In contrast, the Green Revolution fields over wide areas of India became salt deserts, and even the World Bank, which financed many of them, later acknowledged that they had created deserts in the name of making gardens.

It hardly need be added that our meat production is equally destructive and dangerous, not to mention exceedingly cruel. Cancer in chickens, bovine encephalopathy, and illness and death in humans due to various meat-infesting pathogens are taking their heavy toll.

Nevertheless, high-tech agriculture continues to be subsidized by public funds and is justified by misleading "success stories," such as that one U.S. farmer at the turn of the century could feed only four people, while today a single farmer feeds seventy to eighty people. Such statistical distortion flagrantly ignores the army of people and resources producing the chemical herbicides, pesticides and fertilizers, the rapidly obsolete heavy machinery, the fuels and irrigation systems, and the genetically engineered sterile seed that must be bought annually. In fact, the natural farmer at the turn of the century produced ten calories of food energy for every one calorie of energy input and kept his soil and water table healthy, while the present-day farmer puts ten calories of energy into his farm for every one calorie of food he gets out. Meanwhile his land is increasingly impoverished, thus destroying the very basis of his livelihood. High-tech agriculture must be counted as enormously inefficient, energy-wasteful and dangerous to the health of our species, not to mention many others, through its poisoned crops, destruction of soils, and deadly pollution of river and ocean waters .

It is also argued that high-tech agriculture is necessary to produce the sheer volume of food required by today's populations. The case of India above belies this, as do the production figures of restored traditional techniques. In the Philippines, one of the countries where high-tech Green Revolution techniques were pioneered, the restoration of traditional organic rice-growing methods proved superior in quantity of production. Bill Mollison's intensive natural permaculture, which has adapted much indigenous and traditional knowledge, is now taught in over seventy countries and is highly successful, as are French intensive and other concentrated natural agricultures, which create balanced self-restoring ecosystems with high productivity.

A century ago, a British agricultural expert toured India to see how he could best advise Indian farmers to improve their agricultural practices. His conclusion, reported in *The Ecologist* magazine, was that the traditional Indian farmers had more to offer English farmers in the way of advice, because they knew so much about soil composition and health, pest control, water management, crop breeding, and all other aspects of agriculture. They were highly knowledgeable and productive, failing only when they lacked access to natural resources.

The amazing agricultural development and ecologically sound practices of the Incan and pre-Incan Andes exceeded that of any other cultures in history. Most of the world's food today is derived from this Andean agricultural science (consider corn, potatoes, and amaranth grains to begin with). Andean genetic, agricultural, and environmental sciences, if practiced by the world at large, might well make the difference between the survival or extinction of our human species over the next few decades. In 1993, at an International Monetary Fund lecture in Washington D.C., Oswaldo Rivera and Alan Kolata reported on the restoration of the ancient (400–1,000 A.D.) pre-Inca *waru waru,* or chinampa-type agriculture, in the Lake Titicaca altiplano of Peru and Bolivia. After hearing them, I visited this system to see how it increased local annual production from the previous norm of 2.5 tons per hectare to forty tons in only five years, with no chemical fertilizers or pesticides and relatively little work beyond letting water through sluice gates into ditches dug between soil mounds, and planting seeds without plowing. In this system nature creates its own fertilizers, the canals becoming a nutrient sump for nitrogen and phosphorus through colonization by fish, birds, and water plants. The system's automatic irrigation also creates climate control that prevents crops from freezing. The usual crops were varieties of potatoes, grains (including maize, quinoa, and other amaranth grains), and legumes. Now winter wheat, barley, oats, turnips, and other vegetables have been added—even lettuce at an altitude of four thousand meters.

While a handful of books and occasional films on public television document such successes, relatively little is done to convert high-tech agriculture to less destructive, more efficient, and less expensive organic agriculture. Agribusinesses and development banks have so much invested in the short-term profits of this unintelligent, destructive system that they are apparently still blinded to its long-term consequences. The hope that ever-new technologies can solve the problems arising from technology is deeply ingrained in Western culture, and the idea of going back to older methods is anathema to the technologically oriented. Yet a few European countries, such as Denmark and Holland, are now banning chemicals, training professional farm-sitters to give farmers time off, and implementing other sound practices based on traditional agriculture and "upgraded" by appropriate technology. We can only hope that widespread understanding of holarchic biology will lead to far more sustainable practice in producing our all-important food supply.

HARMAN:

You're right, that's a good example. Another example of modern society's ways that conflict with principles of living system is industry itself.

One of the most remarkable stories to come out of the sustainability controversy is that of Karl-Henrik Robèrt, originator of "The Natural Step" which you mentioned earlier. A noted cancer researcher, Dr. Robèrt begins his understanding with cellular biology, because it is the basis for all but the tiniest forms of life. Cells grew and evolved over billions of years through self-sustaining cycles wherein all waste was constantly cycled back to other forms of life. The primary production units are the cells of the green plants, which accomplish photosynthesis. They are unique in their ability to synthesize more structure than is broken down elsewhere in the biosphere. This implies a general requirement for production in the cycles of nature as well as in our societies: Waste products must be recruited to photosynthesis, or recycled within society, or stored away into final deposits.

Nature's toxins have evolved over thousands to millions of years as a part of complex, cyclical, life-giving cycles. They do not break the cyclical pattern of growth, death, and evolution. On the other hand, our man-made poisons, toxins, and chemical and nuclear wastes have no such history. They cannot be taken up and incorporated by the normal metabolic processes of cellular life.

Industrial society has created an accumulation of waste products, garbage, and pollution, disturbing the biosphere, as well as a corresponding decrease in the stocks of natural resources. Furthermore, due to complexity and delay mechanisms, we cannot in general foresee the time limits for the socioeconomic consequences or development of diseases. To continue along this path is not compatible with wealth, nor with human and ecological health.

It has been the genius of "The Natural Step" to ask systemic questions that avoid the technical questions about which scientists will quibble for years to come and thus achieve a consensus permitting cooperative action, whereas technical uncertainties had previously paralyzed the usual political processes. These systemic questions elicit surprising agreement, from Greenpeace and unions to industry and religion. They are asked with regard to CFCs, dioxin, or other man-made substances: *Is this a naturally occurring substance? No. Is it chemically stable? Yes. Does it degrade into harmless substances? No. Does it accumulate in organic or bodily tissues? Yes. Is it possible to predict the acceptable tolerances? No. Can we continue to place dioxin into the environment?* No, not if we want to survive.

The Natural Step program was officially launched in Sweden in 1989. Its activities support a shift away from linear, resource-wasting, toxic-spreading methods of materials handling and manufacturing, toward cyclical, resource-preserving methods. The central strategy is to lend active support to good examples of ecocyclical development in households, companies, and local governments. The program enjoys amazingly broad-based support from major corporations, small businesses, banks and insurance companies, the State Railways, the Church of Sweden, professional networks, and youth groups. It is being emulated in several other northern European countries.

For a truly long-term sustainable economy there are, according to Robèrt, four "non-negotiable" system conditions:

1. Stored mineral deposits: The use of virgin mineral deposits must not exceed the very slow sedimentation processes in nature. In practical terms, this requires an almost full stop to mining.

2. Alien compounds: There must be a phase-out of persistent, non-natural compounds. If the use of such molecules exceeds the slow processes by which Nature destroys them, the principle of matter conservation, together with the scattering tendency (tendency of entropy to increase), will cause accumulation of molecular garbage in the biosphere.

3. Ecosystems: Physical conditions (area and ecological) of nature's diversity and capacity for primary production must be preserved. In practical terms this implies ecologically sustainable farming and forestry, powerful measures to deal with water scarcity, and ceased expansion of the large cities' infrastructures.

4. *Metabolism:* The use of energy and materials must be reduced to within the capacity of the ecosystems to process garbage into new resources. In practical terms this implies a less energy-intensive lifestyle in the Western world, in combination with powerful measures to regulate population growth and to improve the quality of life in the Third World.

What both of these examples (agriculture and industry) represent is the failure of modern human society to co-evolve with its natural environment.

SAHTOURIS:

All creatures and their environments, largely composed of other species, co-evolve by changing themselves and one another. To understand any particular species we must try to understand how its evolution is related to the evolution of its environment. In particular, we can only understand ourselves as humans by trying to understand our co-evolution with our environments over the course of our history.

Imagine the co-evolution of humanity and its ecosystems as though watching a short film: Small groups of humans are seen evolving from their ancestral apes in the dense forests of near-equatorial regions. The climate changes, the forests shrink; we see the evolving humans exchange a life of swinging through trees for one of walking upright on the ground. Groups of them wander in search of food. Slowly they begin making tools and weapons, taming fire, wearing clothes—using the resources of their environment to make things that are of use to them, things that compensate for their lack of fur, sharp claws, and long teeth; things that help them hunt other large animals for food, bone tools, and clothing; things that help them carry, store, and pre-pare food. When their families or tribes grow too large to live together easily, some members bud off to form new tribes.

The human creatures thrive, multiplying and spreading out to follow food and water supplies. Great ice ages push them back toward the equator, but each time the ice thaws, they are lured toward the lush new growth springing up in the wake of the ice. Eventually, food supplies draw them to all the continents. Some remain tribal hunter-gatherers or nomads, while others begin the settling process we call civilization.

In the best climates, groups of them settle to make villages and fields, to keep animals and grow crops, to store food for dry or cold seasons. Villages grow into towns, and towns into larger agricultural societies that transform considerable plots of ground from natural to artificial ecosystems. Living in peace and equality for thousands of years, budding off new colonies as they grow and thus alternating with nomadic existence, they spread over the habitable areas of the world, developing their arts of plant selection, animal husbandry, pottery, painting, and metalwork. Then they are suddenly overrun by other humans, tribes of wan-dering nomads, and hunters from harsher climates, armed with weapons, who take them over and establish a dominance system of males over females—rulers over those ruled. They build king-doms and unite them into empires through warfare. More and more land is taken for human use. The old self-creating, self-balancing ecosystems are destroyed as natural plants are cut or burned and their animals driven off, both replaced by human-bred crops and livestock, as well as by cities of stone, brick, and wood.

Within and between empires, wars are fought and goods traded, building networks of land and sea paths that connect human societies with each other. Along these paths, news, ideas, and stories flow together with people and their products, animals, seeds, and microbes. Sometimes unwittingly, people change whole ecosystems as their seeds or animals take over and drive out the native species. Cities, in which natural land is replaced by man-made buildings and streets, grow up as centers of ideas, inventions, new ways of life. Their crowded conditions also breed disease; plagues sometimes wipe out whole populations.

The borders around kingdoms and empires change; continents are mapped into countries; human populations grow and divide into ever more languages and cultures. The environment has shaped human civilization by drawing it to favorable climates, into fertile river valleys, and along the easiest overland transportation routes. In turn, humans transform the environment ever more to their use. Whole forests are cut for lumber and fuel or burned to clear land for grazing and agriculture. More and more natural land is plowed under by farmers and paved over by builders of cities. Deserts grow larger while more and more species of animals and plants are killed off as humans exploit nature for their own purposes.

Cities crowd more and more people together in artificial environments; raw materials are transported to the city centers from more and more distant places, while the products manufactured from these materials flow outward again toward markets. Crops and animals native to one part of the world are planted and raised in others. Human technology evolves from equipping riding horses and building sailing ships to steamships, jet planes, and spacecraft, from weaving looms to computer industries, from stamping clay tablets to printing presses, from town criers to television. A world once dark by night except for forest fires is lit by a twinkling cobweb of electric lights. A world once silent by night except for the lone cry of a bird or mammal is filled with the sounds of machines and music. Mines and quarries have been dug deep into the Earth and scratched out of its surface, their stone, metal ores, and fossil fuels transformed into human products. Rivers have been dammed up and diverted into unnatural paths, flooding ecosystems behind them, making deserts in front of them, for the sake of the insatiable human demand for electrical power.

The atmosphere, the waterways, the soil, and the oceans become polluted by man-made fertilizers, pesticides, heavy metals, and other waste materials of human production. Yet increased food supplies have exploded humanity itself into such numbers that it seems to humans that they are running out of space—that the Earth is bursting at its seams. Deserts that were created by man are later flooded by man in the hope of extending agriculture; in a few years they dry up again as salt deserts.

Nuclear energy is discovered; two atomic bombs are blown up to deliberately destroy man's own artificial ecosystems; others are blown up just as tests, destroying natural ecosystems, raining fallout from the atmosphere worldwide. Human technology makes the leap into space, and humanity, for the first time ever, sees its exquisitely lovely planet from afar, as a living whole. Humanity suddenly awakens to the recognition of the vast damage it has done to its environment, begins to fear the exhaustion or irreversible pollution of natural waters, fossil fuels, and other supplies, to recognize its power to destroy the whole human world and force the planet into new paths of evolution, to feel the effects of its greenhouse gases in an atmosphere that is growing uncomfortably warm, threatening the end of our species in yet another way we have paved.

It is an impressive scenario, but the saga ends on a frightening note. One species—new, an upstart—has appropriated virtually the entire planet to itself, turning rich and varied ecosystems into fragile monocultures, vast deserts, and choking pollution. Is it a kind of planetary cancer, looking heedlessly to its own expansion at the expense of its own support system? Why is the only species with so much capacity for hindsight and foresight so destructive to itself and its planet? The most obvious feature of human social, political, and economic systems continues to be empire building through dominance: the female half of the species is still largely under the control of and exploited by the male half, most of the Earth's countries are still dominated and exploited by the few most powerful ones, individual countries maintain their own dominance systems of class, caste, and discrimination, the few hiring the many to work for them to bring them wealth and the power for further expansion.

Increasingly, empire building shifts from nations colonize nations to megacorporations that colonize populations around the world for cheap labor and markets most open to their products. Will nations become obsolete? Or will national governments increasingly mobilize their resources for the meager welfare of those dispossessed by labor-downsizing high-tech corporations? In Africa and Asia large dispossessed populations roam frantically in search of water and food; disease is rampant among them. New diseases plague even affluent populations as technological medicines backfire, driving microbes into new genetic configurations, and immune systems falter under the assault of environmental poisons and other stresses. Male fertility is down forty percent worldwide, as other species succumb to widespread human pollution at a higher rate than any previous extinction. Will the human species go extinct? Or will it come to understand cooperative living systems in its darkest hour and transform itself into a global cooperative community that works intensively to restore health to its ecosystems and itself?

HARMAN:

You certainly pose a fundamental problem. The dominant way of thinking about our relationship to the Earth and nature, guided by reductionistic science, is leading the modern world to a non-sustainable future. Can the direction of that cultural evolution be changed rapidly enough to move us toward a positive future, with truly positive relationships with our fellow creatures and the global environment as a whole? Does that involve shifting from physics as the core science to putting living systems in the center of our worldview?

We are coming to the end of this phase of our exploration. I think I'd like to cap it off with some thoughts from Margaret Wheatley. Margaret is a consultant to businesses and organizations, helping them to recognize that it is time to become more self-consciously self-forming. She has put forth some very well-founded principles that make clear the implications of holistic thinking for organizations and societies. These principles may not have been derived explicitly from the "holons within holons" ontological stance we have been taking, but they could have been.

Wheatley's eight principles are (1996):

1. We live in a world in which life wants to happen. In this broad sense, holistic biology is teleological. Not that there are fixed and specific goals, but that creative life seems bent on expressing its creativity beyond measure, and appears purposeful in that sense. We were all influenced, in our growing up, by the accepted Darwinian dogma that the emergence of life on Earth was an accident, and evolution was a series of meaningless accidents and "survival of the fittest." The new view, far more wholesome and more true to total human experience, is that life wants to happen as a community, and we are all part of it.

2. *Organizations and societies are living systems.* And living systems are self-organizing. The new "management" of organizations involves respecting and trusting their self-organizing tendencies. This is also the key to true democracy; in politics we are trying to fumble our way in the same direction.

3. *We live in a universe that is alive, creative, and experimenting all the time to discover what's possible.* We can see that at the largest, or the smallest, levels of scale, whether we're looking at the smallest microbes or out into the galaxies. People are intelligent, creative, adaptive; we seek order, we seek meaning in our lives. When we really start to believe this, it changes how we think about organizing.

4. *It is the natural tendency of life to organize—to seek greater levels of complexity and diversity.* Life seeks to affiliate with other life, and as it does it makes more possibilities available. It seeks to create patterns, structures, organization, without preplanned directive leadership.

5. *Life uses messes to get to well-ordered solutions.* What may appear to be messy and inefficient, upon deeper perception looks like life experimenting—discovering what is possible. In the re-creation of ecosystems, for example, it takes a lot of messes before the discovery of what really works for multiple species, but the direction is always toward order.

6. *Life is intent on finding what works, not what's right.* When you look around, you see life tinkering, experimenting, playing. Playfulness comes creatively into human relationships, where the task of any moment is to find something that works, and not be ego-attached to finding the "right" answer.

7. *Life creates more possibilities as it engages with opportunities.* We sometimes hear it said that some aspect of life presents a "narrow window of opportunity." This is never true; living systems don't work that way. Every time we try to make something work, we are creating more possibilities within the system—we open many different "windows of opportunity." If a particular opportunity is not fulfilled, there are always many others to engage with. Each path of opportunity leads to its own pattern of order.

8. *Life organizes around identity.* Out of all this blooming, buzzing confusion of life, we look for patterns and information that are meaningful to us in some way, given who we think we are. Life organizes spontaneously and creatively around a self; all of life has this subjective dimension. Consciousness is at work in everything, forming itself into different identifiable beings.

Margaret Wheatley's insights are, to me, profound. They express in another way the "story" I think we will tell ourselves when a more holistic biology has come to complement the extremely effective technology-focused science that presently prevails.

And now it is time to invite you, dear reader, to actively enter the dialogue. We, Willis and Elisabet, have let you in on our more intimate thinking about some important issues. But that's only the beginning. The questions we have raised about appropriate science, biology as the core science, useful metaphors for a more holistic biology, implications for the "story" our culture tells itself about the emergence and evolution of living systems, further implications for dealing with self-organizing organizations and societies—these questions are important. They are not just important to scientists and philosophers—they are critical questions for every citizen of this transforming society.

Some of you will get involved with conversations about these issues; some will write about them; some will involve yourselves in a dialogue on the Internet. We hope that what we have done here will be a helpful stimulus to all of that. Vive la conversation!

References

Abram, David (1996), *Spell of the Sensuous*. New York: Pantheon Books.

Assagioli, Roberto (1965), *Psychosynthesis: A Manual of Principles and Techniques*. New York: Viking.

Augros, Robert and George Stanciu (1987), *The New Biology: Discovering the Wisdom in Nature*. Boston: Shambhala, New Science Library.

Baldwin, William (1993), *Spirit Releasement Therapy*. Falls Church, Virginia: Human Potential Foundation.

Barfield, Owen (1982), "The Evolution Complex," *Towards*, vol. 2, no. 2, Spring 1982, pp. 6-16.

Bateson, Gregory (1980), *Mind and Nature: A Necessary Unity*. New York: Bantam Books.

Beloff, John (1977), "Psi Phenomena: Causal Versus Acausal Interpretation." *Jour. Soc. Psychical Research* vol. 49, no. 773; Sept. 1977.

Bentov, Itzhak (1977), *Stalking the Wild Pendulum,* New York: E.P. Dutton.

Bortoft, Henri (1996), *The Wholeness of Nature: Goethe's Science of Conscious Participation in Nature.* Hudson, NY: Lindisfarne Press.

Brown, Courtney (1996), *Cosmic Voyage.* New York: Dutton.

Cairns, John, Julie Overbaugh and Stephan Miller (1988), "The Origin of Mutants." *Nature,* 335 (September 8): 142-145.

Cahete, Greg (1994), *Look to the Mountain: An Ecology of Indigenous Education.* Durange, Colorado: Kivaki Press.

Campbell, Donald T. (1974), "Downward Causation in Hierarchically Organized Biological Systems," *Studies in the Philosophy of Biology,* eds. F. Ayala & T. Dobzhansky. Berkeley: University of California Press.

Davidson, John (1992), *Natural Creation or Natural Selection?* Shaftesbury, Dorset: Element Books.

Denton, Michael (1985), *Evolution: A Theory in Crisis.* Bethesda, Maryland: Adler and Adler.

Depew, David & Bruce Weber, eds. (1985), *Evolution at a Crossroads: The New Biology and the New Philosophy of Science.* Cambridge: MIT Press.

Dobzhansky, Theodosius (1967), *The Biology of Ultimate Concern.* New York: New American Library.

Easlea, Brian (1983), *Fathering the Unthinkable: Masculinity, Scientists and the Nuclear Arms Race.* London: Pluto Press.

REFERENCES

Edelman, Gerald M. (1992), *Bright Air, Brilliant Fire: On the Matter of the Mind* New York: Basic Books.

Eisenbud, Jule (1983), *Parapsychology and the Unconscious.* Berkeley, California: North Atlantic Books.

Eldredge, Niles (1987), *Life Pulse.* London: Facts On File Publications.

Endler, John A. (1986), *Natural Selection in the Wild.* Princeton, New Jersey: Princeton University Press.

Fleischaker, Gail R. (1990), "Origins of Life: An Operational Definition." *Origins of Life and Evolution of the Biosphere* **20:** 127-137.

Friedman, Norman (1994), *Bridging Science and Spirit.* St. Louis, Missouri: Living Lake Books.

Fukuoka, Masanobu (1987), *The Road Back to Nature: Regaining the Paradise Lost.* Japan Publishers, Inc.

Goodwin, Brian (1994a), *How the Leopard Changed Its Spots: The Evolution of Complexity.* London: Charles Scribner's and Sons.

Goodwin, Brian (1994b), "Toward a Science of Qualities," in *New Metaphysical Foundations of Modern Science,* W. Harman ed. Sausalito, California: Institute of Noetic Sciences.

Gould, Stephen Jay (1989), *Wonderful Life: The Burgess Shale and the Nature of History.* New York: W. W. Norton.

Hall, Barry G. (1988), "Adaptive Evolution That Requires Multiple Spontaneous Mutations: I. Mutations Involving an Insertion Sequence." *Genetics,* 120 (December): 887-897.

Hastings, Arthur (1991), *Tongues of Men and Angels.* Fort Worth, Texas: Holt, Rinehart and Winston.

Hawken, Paul (1993), *The Ecology of Commerce: A Declaration of Sustainability.* New York: Harper Business.

Hefner, Philip (1993), *The Human Factor: Evolution, Culture, and Religion.* Minneapolis: Fortress Press.

Henderson, Hazel (1996), *Building a Win-Win World.* San Francisco: Berrett-Koehler.

Heywood, Rosalind (1974), *Beyond the Reach of Sense: An Inquiry into Extra-Sensory Perception.* New York: E. P. Dutton.

Ho, Mae-Wan and P.T. Saunders, eds. (1984), *Beyond Darwinism: Introduction to the New Evolutionary Paradigm.* London: Academic Press.

Ho, Mae-Wan and S.W. Fox, eds. (1988), *Evolutionary Processes and Metaphors.* London: Wiley.

Horgan, John (1996), *The End of Science: Facing the Limits of Knowledge in the Twilight of the Scientific Age.* New York: Addison-Wesley.

Hoyle, Fred (1983), *The Intelligent Universe,* London, Michael Joseph Ltd.

Hunt, Valerie Hunt (1995), *Infinite Mind: the Science of Human Vibrations.* Malibu, California: Malibu Publishing Co.

Inglis, Brian (1992), *Natural and Supernatural: A History of the Paranormal.* Bridport, Dorset: Prism Press.

Jahn, Robert and Brenda Dunne (1987), *Margins of Reality: The Role of Consciousness in the Physical World.* New York: Harcourt Brace Jovanovich.

James, William (1912), *Essays in Radical Empiricism.* New York: Longmans, Green and Co.

Jantsch, Erich (1980), *The Self-Organizing Universe.* Oxford: Pergamon Press.

Jantsch, Erich & Waddington, C. H. (1976), *Evolution and Consciousness.* Reading, Massachusetts: Addison-Wesley.

Johnson, Raynor C. (1957), *Nurslings of Immortality.* New York: Harper and Brothers. (Contains a summary of the conclusions of F. W. H. Myers' books *The Road to Immortality* and *Beyond Human Personality.*)

Jung, C. G. and Pauli, W. (1955), *The Interpretation and Nature of the Psyche,* translated by R. F. C. Hull and P. Silz. New York: Pantheon.

Kaku, Michio (1994), *Hyperspace.* Oxford University Press.

Korten, David (1995), *When Corporations Rule the World.* San Francisco: Berrett-Koehler.

Korten, David (1996, March/April), "Economies of Meaning." *Tikkun:* 17-19.

Kuhn, Thomas (1970), *The Structure of Scientific Revolutions,* 2nd ed. Chicago: University of Chicago Press.

Lapo, A. V. (1982), *Traces of Bygone Biospheres.* Moscow: Mir Publishers.

Laszlo, Ervin (1996), *Evolution: Foundations of a General Theory,* Alfonso Montuori, ed. NJ: Hampton Press.

Leakey, Richard, and Roger Lewin (1995), *The Sixth Extinction: Patterns of Life and the Future of Humankind.* New York: Doubleday.

Levins, Richard and Richard Lewontin (1985), *The Dialectical Biologist.* Cambridge: Harvard University Press.

Lipton, Bruce (1993), "The Biology of Consciousness," *Proc. of the Int'l Assoc. of New Sciences.*, Ft. Collins, Colorado.

Lovelock, James (1979), *Gaia: A New Look at Life on Earth.* New York: Oxford University Press.

Lovelock, James (1988), *The Ages of Gaia: A Biography of Our Living Earth.* New York: W.W. Norton.

Lucas, Winafred (1993), *Regression Therapy: A Handbook for Professionals.* Crest Park, California: Deep Forest Press.

Macy, Mark and Pat Kubis (1995), *Conversations Beyond the Light.* Boulder, Colorado: Griffin Publishing. Up to date information on "instrumental transcommunication" (ITC) available from Mark Macy, Continuing Life Research, P.O. Box 11036, Boulder, CO 80301.

Margulis, Lynn (1991), "Lynn Margulis: Science's Unruly Earth Mother," in Science, **252,** pp. 378-381.

Margulis, Lynn, and Dorion Sagan (1986), *Microcosmos: Four Billion Years of Evolution from Our Microbial Ancestors.* New York: Simon & Schuster.

Margulis, Lynn, and Dorion Sagan (1995), *What is Life.* New York: Simon & Schuster.

Maturana, Humberto and Francisco Varela (1987), *The Tree of Knowledge: The Biological Roots of Human Understanding.* Boston: New Science Library.

Mayr, Ernst (1988), *Toward a New Philosophy of Biology: Observations of an Evolutionist.* Cambridge, Mass.: Harvard University Press.

McClintock, Barbara (1984), "The significance of responses of the genome to challenge," *Science,* 226, 792-801.

Milla Villena, Carlos (1983), *Genesis de la Cultura Andina.* Lima, Peru: Fondo Editorial C.A.P. Coleccion Bienal.

Mitchell, Edgar D. (1974), *Psychic Exploration.* New York: G. P. Putnam's Sons.

Pankow, Walter (1976), "Openness as Self-Transcendence," in Jantsch, E. & Waddington, C. H., *Evolution and Consciousness.* Reading, Massachusetts: Addison-Wesley.

Peat, F. David (1987), *Synchronicity: The Bridge Between Matter and Mind.* New York: Bantam.

Pollard, Jeffrey (1988), "New Genetic Mechanisms and Their Implications for the Formation of New Species," in Ho, Mae-Wan and Fox, Sidney, eds., *Evolutionary Processes and Metaphors.* John Wiley & Sons.

Prigogine, Ilya and Stengers, Isabelle (1984), *Order out of Chaos: Man's New Dialogue with Nature.* New York: Bantam.

Puthoff, Harold (1990), "Everything for Nothing." *New Scientist* 28 July.

Quine, W.V.O. (1960), *Word and Object.* Cambridge, Massachusetts: MIT Press.

Rabinovitch, Jonas and Josef Leitman (1996), "Urban Planning in Curitiba." *Scientific American,* April 1996.

Sahtouris, Elisabet (1991), "Beautiful Bulrushes, Remarkable Reeds: The Water Reclamation Miracles of Dr. Kaethe Seidel." Website: www.ratical.com/lifeweb.

Sahtouris, Elisabet (1996), *EarthDance*. Santa Barbara: Metalog. Website: www.ratical. com/lifeweb.

Sahtouris, Elisabet (1997), "The Biology of Globalization," in *Perspectives on Business and Global Change,* September, 1997.

Salk, Jonas and Jonathan Salk (1981), *World Population and Human Values: A New Reality.* New York: Harper & Row.

Salk, Jonas (1983), "A Conversation with Jonas Salk," in *Psychology Today,* March 1983.

Scotton, Bruce, Allen Chinen and John Battista (1996), *Textbook of Transpersonal Psychiatry and Psychology.* New York: Basic Books.

Sheldrake, Rupert (1981), *A New Science of Life: The Hypothesis of Formative Causation.* Los Angeles: J.P. Tarcher.

Sheldrake, Rupert (1988), *The Presence of the Past: Morphic Resonance and the Habits of Nature.* New York: Times Books.

Shiva, Vandana (1988), *Staying Alive.* London: Zed Press.

Shiva, Vandana (1989), *The Violence of the Green Revolution.* Dehra Dun, India.

Sinnott, Edmund W. (1955), *The Biology of the Spirit.* New York: Viking Press.

Sonea, S. and M. Panisset (1983), *A New Bacteriology.* Boston: Jones & Bartlett.

Stevenson, Ian (1987), *Children Who Remember Previous Lives.* Charlottesville: University Press of Virginia.

Tart, Charles (1975a), *States of Consciousness.* New York: Dutton.

Tart, Charles, ed., (1975b), *Transpersonal Psychologies.* New York: Harper and Row.

REFERENCES

Teilhard de Chardin, Pierre (1959), *The Phenomenon of Man.*
London: William Collins.

Temin, H. M., and Engels, W. (1984), "Movable Genetic Elements
and Evolution," in J. W. Pollard, ed., *Evolutionary Theory: Paths
into the Future.* Chichester: John Wiley & Sons.

Thomas, Lewis (1975), *Lives of a Cell: Notes of a Biology Watcher.*
New York: Bantam.

Tompkins, Peter, and Christopher Bird (1973), *The Secret Life of
Plants.* New York: Harper and Row.

Underwood, Paula (1991), *Three Strands in the Braid: A Guide for
Enablers of Learning.* San Anselmo, California: A Tribe of Two
Press.

Velmans, Max (1993), "A Reflexive Science of Consciousness," in
Experimental and Theoretical Studies of Consciousness, Ciba
Foundation Symposium No. 174. Chichester: Wiley.

Waddington, C. H. (1961), *The Nature of Life.* London: Allen &
Unwin.

Wald, George (1987), "The Cosmology of Life and Mind," in
Singh, T. D. and Ravi Gomitam, eds., *Synthesis of Science and
Religion: Critical Essays and Dialogues.* San Francisco: The
Bhaktivedanta Institute.

Waldrop, M. Mitchell (1992), *Complexity.* New York: Simon &
Schuster.

Walsh, Roger and Frances Vaughan (1993), "The Riddle of
Consciousness." Unpublished paper.

Watzlawick, Paul, (1984), *The Invented Reality.* New York: W. W.
Norton.

Wesson, Robert G. (1989), *Cosmos and Metacosmos*. La Salle, Illinois: Open Court.

Wesson, Robert G. (1991), *Beyond Natural Selection*. Cambridge, MA: MIT Press.

Wheatley, Margaret, and Myron Kellner-Rogers (1996), *A Simpler Way*. San Francisco: Berrett-Koehler.

Wilber, Ken (1993), "The Great Chain of Being." *Journal of Humanistic Psychology,* Vol. 33 no. 3, Summer, pp. 52-65.

Wilber, Ken (1995), *A Brief History of Everything*. Boston: Shambhala.

Wyller, Arne (1996), *The Planetary Mind*. Aspen, Colorado: MacMurray & Beck, Inc.

Zukav, Gary (1991), *Science and Spirit*. Proceedings of the 3rd International Forum on New Science, Fort Collins, Colorado September, 1991.

Index

About the Institute of Noetic Sciences

We live in a time of rapid environmental, social, scientific, and cultural change. In this era of globalization and transformative change, we recognize that many of our most fundamental assumptions about human nature are being called into question. The Institute of Noetic Sciences, founded in 1973, is a research foundation, an educational institution, and a membership organization comprised of individuals who are committed to developing our understanding of human consciousness through scientific inquiry, spiritual understanding, and psychological well-being. The word *noetic* is derived from the Greek word for mind, intelligence, or transcendental experience. Therefore "noetic sciences" are those that study the mind, and its diverse ways of knowing, in a truly interdisciplinary fashion.

As a research foundation...

Our intention is to support innovative researchers and promising new directions of inquiry. We conduct original research, and provide research contracts for leading-edge scientific and scholarly research. In addition, we serve as a networking center and as an information clearinghouse.

As an education institution, we publish a quarterly journal, a quarterly member magazine, and program-relevant monographs, videos and books. The Institute also underwrites *New Dimensions Radio,* a weekly program on national public radio, and is affiliated with the *Thinking Allowed* television program and the Hartley Film Foundation.

As a membership organization, we offer opportunities for individuals to integrate scientific and scholarly research with their own experiences. Members participate in study groups, conferences and meetings, and contribute to research concerned with members' perceptions and experiences in a changing world. A travel program provides opportunities to explore diverse cultures around the world.

If you would like to support our research, or for more information about the Institute, please contact us at:

The Institute of Noetic Sciences
475 Gate Five Road, Suite 300-Research
Sausalito CA 94965
phone 415-331-5650 • fax 415-331-5673
e-mail: research@noetic.org

About the Authors

The late **WILLIS HARMAN, PH.D.** was the president of the Institute of Noetic Sciences, a non-profit research, education, and membership organization founded in 1973. Its mission is to expand knowledge of the nature and potentials of the mind, and to apply that knowledge to the advancement of  health and well-being for humankind. Dr. Harman directed the Institute's Causality Program, an inquiry into the fundamental assumptions of contemporary science, and an exploration of anomalous phenomena that point toward the need for an expanded or revised science. The Causality group explored what that science might look like, with special emphasis on developing ideas for a science that could include consciousness as a causal factor. He was a founding board member of the World Business Academy and a world-renowned lecturer and futurist. Dr. Harman's recent books include *New Metaphysical Foundations of Modern Science, Global Mind Change* (a new edition is forthcoming in 1998), *An Incomplete Guide to the Future, Higher Creativity,* and *Changing Images of Man.*

ELISABET SAHTOURIS, PH.D. is an American/Green evolutionary biologist, ecologist, and futurist. She is a United National consultant on indigenous peoples, and a founding member of the Worldwide Indigenous Science Network who has worked at the Earth Summit in Rio de Janeiro. She has served as a member of the Earth Parliament and the Women's International Policy Action Committee on Environment and Development. Dr. Sahtouris is also an advisor to the Institute for Sustainable Development and Alternative Futures, Global Education Associations, and the Earth Restoration Corps. She published *Gaia: The Human Journey from Chaos to Cosmos,* (newly retitled *Earthdance*), which has been printed in eight languages. Dr. Sahtouris has lectured widely in Europe and in Central, South, and North America. She has appeared on television and on radio, and has been published in a wide variety of journals.